D0917052

GROUP THEORY MADE EASY FOR SCIENTISTS AND ENGINEERS

GROUP THEORY MADE EASY FOR SCIENTISTS AND ENGINEERS

NYAYAPATHI V.V.J. SWAMY

MARK A. SAMUEL

Oklahoma State University

A WILEY-INTERSCIENCE PUBLICATION

JOHN WILEY & SONS

NEW YORK
CHICHESTER
BRISBANE
TORONTO

Library of Congress Cataloging in Publication Data

Swamy, Nyayapathi Venkata Vykuntha Jagannadha.
  Group theory made easy for scientists and engineers.

  "A Wiley-Interscience publication."
  Bibliography: p.
  1. Groups, Theory of. I. Samuel, Mark A., joint author. II. Title.

QA171.S88   512'.22   78-11733
ISBN 0-471-05128-4

Printed in the United States of America

10 9 8 7 6 5 4 3 2 1

PREFACE

One might be tempted to ask "Yet another book on Group Theory?" This book is different, however, in that it aims to help the advanced undergraduate, the beginning graduate student, or the industrial researcher who would like to be familiar with the tools of symmetry without having to wade through elaborate mathematical proofs. Extensive theory to be found in most texts on group theory has been avoided; instead, attention is concentrated on discussion of illustrative problems. The choice of material has been dictated by the experience gained in teaching group theory courses, and the areas covered include atomic physics, nuclear physics, particle physics, solid-state physics, and molecular physics. We believe this is a useful addition to the existing literature, one that seeks to supplement rather than duplicate other treatments of the subject, and one that students of mathematics, chemistry, or engineering will also find useful.

The informal approach in this book is motivated by the desire to acquaint the uninitiated with the fundamentals of Group Theory. Chapters 1 - 4 go over the bare essentials and one who needs a quick grasp of the tools can afford to omit Chapters 5 - 8. Each of the latter chapters is meant to familiarize the student with the applications in different fields at the level of background preparation. The book as a whole is designed for a one semester course for students who had courses in Calculus, Elementary Linear Algebra and Modern Physics or Introductory Quantum Mechanics, although Chapters 1 - 4 can be gone through even without the last mentioned preparation.

We would like to thank Ms. Janet Sallee for her diligent typing of the manuscript. We are thankful to Eugene Chaffin and Pat Miller for their collaboration on the Appendixes.

NYAYAPATHI V. V. J. SWAMY

MARK A. SAMUEL

Stillwater, Oklahoma

December 19, 1978

CONTENTS

GROUP THEORY MADE EASY FOR SCIENTISTS AND ENGINEERS

CHAPTER ONE

MATRICES, ALGEBRAS

MATRICES

We will briefly review some properties of square matrices that are helpful in understanding group representations. In the following matrix

$$U = \begin{pmatrix} -\frac{1}{\sqrt{2}} & 0 & \frac{1}{\sqrt{2}} \\ -\frac{i}{\sqrt{2}} & 0 & -\frac{i}{\sqrt{2}} \\ 0 & 1 & 0 \end{pmatrix}$$

we know that the determinant of the matrix can be expanded using the elements in any row:

$$-\frac{1}{\sqrt{2}} \begin{vmatrix} 0 & -\frac{i}{\sqrt{2}} \\ 1 & 0 \end{vmatrix} + 0 \begin{vmatrix} -\frac{i}{\sqrt{2}} & -\frac{i}{\sqrt{2}} \\ 0 & 0 \end{vmatrix} + \frac{1}{\sqrt{2}} \begin{vmatrix} -\frac{i}{\sqrt{2}} & 0 \\ 0 & 1 \end{vmatrix}$$

$$\det U = -\frac{1}{\sqrt{2}} \left(\frac{i}{\sqrt{2}}\right) + 0(0) + \frac{1}{\sqrt{2}} \left(-\frac{i}{\sqrt{2}}\right) = -i$$

$$-\frac{i}{\sqrt{2}} \left(\frac{1}{\sqrt{2}}\right) + 0(0) - \frac{i}{\sqrt{2}} \left(\frac{1}{\sqrt{2}}\right) = -i$$

$$0(0) + 1(-i) + 0(0) = -i \qquad (1)$$

Defining a *reduced cofactor* as the cofactor multiplying the element of each row in the expansion of the determinant divided by the value of the determinant, we obtain a corresponding matrix of reduced cofactors, the so-called *contragredient matrix*

$$\begin{pmatrix} -\frac{1}{\sqrt{2}} & 0 & \frac{1}{\sqrt{2}} \\ \frac{i}{\sqrt{2}} & 0 & \frac{i}{\sqrt{2}} \\ 0 & 1 & 0 \end{pmatrix} \tag{2}$$

The transpose of this contragredient matrix is the *inverse of U*. Thus

$$U^{-1} = \begin{pmatrix} -\frac{1}{\sqrt{2}} & \frac{i}{\sqrt{2}} & 0 \\ 0 & 0 & 1 \\ \frac{1}{\sqrt{2}} & \frac{i}{\sqrt{2}} & 0 \end{pmatrix} \tag{3}$$

$$U\,U^{-1} = U^{-1}\,U = 1$$

where 1 is the unit matrix, $\begin{pmatrix} 1 & 0 & 0 \\ 0 & 1 & 0 \\ 0 & 0 & 1 \end{pmatrix}$. If each element of a matrix is replaced by its complex conjugate and if the rows and columns are then interchanged (transposed), the result is called the *adjoint matrix*

$$U^{+} = \begin{pmatrix} -\frac{1}{\sqrt{2}} & \frac{i}{\sqrt{2}} & 0 \\ 0 & 0 & 1 \\ \frac{1}{\sqrt{2}} & \frac{i}{\sqrt{2}} & 0 \end{pmatrix} \tag{4}$$

We notice in the above instance, however, U^{+} is the same as U^{-1}. Such matrices are known as *unitary matrices*. In this unitary matrix U any two rows, treated as vectors, are orthogonal, and each row normalizes to unity:

$$\left(\frac{i}{\sqrt{2}},\ 0,\ \frac{i}{\sqrt{2}}\right)\begin{pmatrix} -\frac{1}{\sqrt{2}} \\ 0 \\ \frac{1}{\sqrt{2}} \end{pmatrix} = -\frac{i}{2} + 0 + \frac{i}{2} = 0 \qquad \text{(orthogonality of rows 2 and 1)}$$

$$\left(\frac{i}{\sqrt{2}},\ 0,\ \frac{i}{\sqrt{2}}\right)\begin{pmatrix} -\frac{i}{\sqrt{2}} \\ 0 \\ -\frac{i}{\sqrt{2}} \end{pmatrix} = \frac{1}{2} + 0 + \frac{1}{2} = 1 \qquad \text{(row 2 normalized).} \tag{5}$$

The columns have similar properties. If a matrix is *equal* to its adjoint it is called a *Hermitean matrix*. For instance,

$$H = \begin{pmatrix} -\frac{1}{\sqrt{2}} & 0 & \frac{1}{\sqrt{2}} \\ 0 & 1 & \frac{i}{\sqrt{2}} \\ \frac{1}{\sqrt{2}} & -\frac{i}{\sqrt{2}} & 0 \end{pmatrix} = H^{+} \tag{6}$$

Let A and B be two 2 x 2 matrices

$$A = \begin{pmatrix} a & b \\ c & d \end{pmatrix}, \quad B = \begin{pmatrix} e & f \\ g & h \end{pmatrix} \tag{7}$$

Two 4 x 4 matrices are derived from these two matrices, the direct sum $A \oplus B$ and the direct product $A \otimes B$

$$A \oplus B = \begin{pmatrix} a & b & 0 & 0 \\ c & d & 0 & 0 \\ 0 & 0 & e & f \\ 0 & 0 & g & h \end{pmatrix}, \quad B \oplus A = \begin{pmatrix} e & f & 0 & 0 \\ g & h & 0 & 0 \\ 0 & 0 & a & b \\ 0 & 0 & c & d \end{pmatrix}$$

$$A \otimes B = \begin{pmatrix} aB & bB \\ cB & dB \end{pmatrix} = \begin{pmatrix} ae & af & be & bf \\ ag & ah & bg & bh \\ ce & cf & de & df \\ cg & ch & dg & dh \end{pmatrix}$$

$$B \otimes A = \begin{pmatrix} ea & eb & fa & fb \\ ec & ed & fc & fd \\ ga & gb & ha & hb \\ gc & gd & hc & hd \end{pmatrix} . \tag{8}$$

The trace of either direct product matrix (the sum of diagonal elements) is seen to be (a + d) (e + h), which is the product of the traces of the two factors in the direct product.

$U\,H'\,U^{-1}$ is called a similarity transformation of a matrix H', U being the transforming matrix. Matrix multiplication shows

$$\underset{U}{\begin{pmatrix} -\frac{1}{\sqrt{2}} & 0 & \frac{1}{\sqrt{2}} \\ -\frac{i}{\sqrt{2}} & 0 & -\frac{i}{\sqrt{2}} \\ 0 & 1 & 0 \end{pmatrix}} \underset{H'}{\begin{pmatrix} 0 & 0 & -1 \\ 0 & 2 & 0 \\ -1 & 0 & 0 \end{pmatrix}} \underset{U^{-1}}{\begin{pmatrix} -\frac{1}{\sqrt{2}} & \frac{i}{\sqrt{2}} & 0 \\ 0 & 0 & 1 \\ \frac{1}{\sqrt{2}} & \frac{i}{\sqrt{2}} & 0 \end{pmatrix}} = \underset{\lambda_i \delta_{ij}}{\begin{pmatrix} 1 & 0 & 0 \\ 0 & -1 & 0 \\ 0 & 0 & 2 \end{pmatrix}}$$

or, symbolically,

$$U\,H'U^{-1} = \lambda_i \delta_{ij} \tag{9}$$

where H' is the matrix $\begin{pmatrix} 0 & 0 & -1 \\ 0 & 2 & 0 \\ -1 & 0 & 0 \end{pmatrix}$

and $\lambda_1 = 1$, $\lambda_2 = -1$, $\lambda_3 = 2$. The Kronecker symbol δ_{ij} has the usual meaning

$$\begin{aligned} \delta_{ij} &= 1 \qquad i = j \\ &= 0 \qquad i \neq j \end{aligned} \tag{10}$$

U is said to diagonalize H' through a similarity transformation, and the diagonal elements are the *eigenvalues* of H'. Notice the trace of the diagonal matrix, that is, the sum of eigenvalues is 1 - 1 + 2 = 2 which is also the trace of H'. This illustrates the well-known theorem, of importance in group theory, that the trace of a matrix is invariant to a similarity transformation. The general procedure for diagonalizing any matrix is to be found in standard texts, for instance D. E. Littlewood (1970) *A University Algebra*.

ALGEBRAS

An algebra in which the associative law of multiplication is valid is called a *linear associative algebra*. The set of all n x n square matrices (matrix elements complex numbers) forms an algebra of order n^2, called the *total matrix algebra*. The basis elements e_{ij} of such an algebra satisfy

$$e_{ij}\, e_{kl} = 0 \qquad j \neq k$$

$$e_{ij}\, e_{jl} = e_{il} \tag{11}$$

For n = 2 these are

$$e_{11} = \begin{pmatrix} 1 & 0 \\ 0 & 0 \end{pmatrix}, \; e_{12} = \begin{pmatrix} 0 & 1 \\ 0 & 0 \end{pmatrix}, \; e_{21} = \begin{pmatrix} 0 & 0 \\ 1 & 0 \end{pmatrix}, \; e_{22} = \begin{pmatrix} 0 & 0 \\ 0 & 1 \end{pmatrix}$$

In other words, the general basis element matrix e_{ij} has 1 in the i-th row and j-th column and zeroes elsewhere.

Several interesting results follow when the basis elements of an algebra are isomorphic to the elements of a group; the multiplication table of the algebra will in this case also be the Cayley table for the group. We illustrate the important properties of such algebras with the help of two finite groups: the Abelian cyclic group C_4 and the group C_{3v} (isomorphic to S_3 or D_3).

We assume the elements of C_4 to be isomorphically represented by basis elements e_i such that E (identity element of the group) $\rightarrow$ e_1, $C_4 \rightarrow e_2$, $C_4^2 \rightarrow e_3$, $C_4^3 \rightarrow e_4$. It is obvious that the constants of multiplication in the algebra are all either 1 or 0 because of the fundamental group property. The common multiplication table is

	e_1	e_2	e_3	e_4
e_1	e_1	e_2	e_3	e_4
e_2	e_2	e_3	e_4	e_1
e_3	e_3	e_4	e_1	e_2
e_4	e_4	e_1	e_2	e_3

$e_i\, e_j = \gamma_{ijk}\, e_k \qquad \gamma_{ijk} = 1 \text{ or } 0$

A general element of the algebra is $X = \sum_{i=1}^{4} x_i e_i$ where x_i are the chosen complex numbers. We notice that there also exists an element of the algebra $e_1 = \sum_{i=1}^{4} \varepsilon_i e_i$, where $\varepsilon_1 = 1$, $\varepsilon_2 = 0 = \varepsilon_3 = \varepsilon_4$, which commutes with every element of the algebra,

$$e_1 x = x\, e_1 = x \tag{12}$$

e_1 is called the *modulus of the algebra*. Furthermore, $e_1^2 = e_1$, and for this reason e_1 is said to be an *idempotent*. The product of no other element with e_1 vanishes; hence e_1 is also called the *principal idempotent*. From Eq. (12) we see

$$\begin{aligned} x^2 = (x_1{}^2 + x_3{}^2 + 2x_2x_4)e_1 &+ (2x_1x_2 + 2x_2x_4)e_2 \\ &+ (x_2{}^2 + 2x_1x_3 + x_4{}^2)e_3 \\ &+ (2x_1x_4 + 2x_2x_3)e_4 \end{aligned} \tag{13}$$

and this cannot be 0 unless all x_i are 0. An element x of an algebra is said to be nilpotent if $x^n = 0$ for some integer n. This algebra is thus not nilpotent and contains an idempotent element. This is a particular application of a theorem that says every algebra that is not nilpotent contains an idempotent element.

We now ask the question: Given x, can we find another element of the algebra $y = \sum_i y_i e_i$ such that it satisfies the equation

$$xy = \omega y$$

or

$$(x-\omega)y = 0 \qquad (14)$$

where ω is a number. If we carry out the multiplication x y and collect coefficients of e_i, this equation takes the form

$$[\ \ ]e_1 + [\ \ ]e_2 + [\ \ ]e_3 + [\ \ ]e_4 = 0 \qquad (15)$$

Since e_i, by definition, are nonzero, each coefficient in Eq. (15) must vanish, and we have four linear equation in y_i, which can be written in matrix form

$$\begin{pmatrix} x_1-\omega & x_4 & x_3 & x_2 \\ x_2 & x_1-\omega & x_4 & x_3 \\ x_3 & x_2 & x_1-\omega & x_4 \\ x_4 & x_3 & x_2 & x_1-\omega \end{pmatrix} \begin{pmatrix} y_1 \\ y_2 \\ y_3 \\ y_4 \end{pmatrix} = 0 \qquad (16)$$

As is well known, for a nontrivial solution the determinant of this matrix $\Delta(x)$, a polynomial in ω of degree 4, has to vanish. $\Delta(x)$ is called the *characteristic determinant* and $\Delta(x) = 0$ the *characteristic equation*.

In the case of a semisimple algebra it can be shown that the characteristic equation can be reduced, by a similarity transformation of the matrix equation, to a product of factors that cannot further be reduced. In our example the unitary matrix

$$U = \begin{pmatrix} \frac{1}{2} & \frac{1}{2} & \frac{1}{2} & \frac{1}{2} \\ \frac{1}{2} & \frac{i}{2} & -\frac{1}{2} & -\frac{i}{2} \\ \frac{1}{2} & -\frac{1}{2} & \frac{1}{2} & -\frac{1}{2} \\ \frac{1}{2} & -\frac{i}{2} & -\frac{1}{2} & \frac{i}{2} \end{pmatrix} \qquad (17)$$

reduces the matrix equation to

$$U[(x-\omega y)y]U^{-1} = 0 \tag{18}$$

$$\begin{pmatrix} (x_1+x_2+x_3+x_4-\omega) & 0 & 0 & 0 \\ 0 & (x_1+ix_2-x_3-ix_4-\omega) & 0 & 0 \\ 0 & 0 & (x_1-x_2+x_3-x_4-\omega) & 0 \\ 0 & 0 & 0 & (x_o-ix_2-x_3+ix_4-\omega) \end{pmatrix} = 0 \tag{19}$$

The characteristic equation is thus reduced to four linear factors, each of the first degree in x_i

$$(x_1+x_2+x_3+x_4-\omega)\ (x_1+ix_2-x_3-ix_4-\omega)\ (x_1-x_2+x_3-x_4-\omega)$$
$$(x_1-ix_2-x_3+ix_4-\omega) = 0 \tag{20}$$

The number of irreducible factors being 4 is a consequence of there being <u>four</u> elements of the algebra (in this case, all the elements!) that <u>commute</u> with every element of the algebra. That each factor is linear in the numbers x_i and raised to the first power is related to the fact that

$$\Sigma_\mu (n_\mu \times n_\mu) = 4 = \text{order of the algebra}$$

where n_μ is the degree of the x_i in each factor, and also the power of that factor in the product. This indirectly verifies the theorem that a semisimple algebra is equivalent to a direct sum of total matrix algebras, here a sum of four 1 x 1 matrix algebras. Of significance is the fact that the C_4 group has <u>four</u> classes. If, for instance, we choose $x_1 = x_2 = x_3 = 0$ and $x_1 = 1$ as the numbers defining the element x, then the roots of the characteristic equation will be

$$\omega = 1, i, -1, -i$$

These are the one-dimensional matrix algebras into which the given algebra decomposes. These numbers are familiar from the character table of C_4!

A more illuminating example is the algebra of order 6 with basis elements $e_1, e_2, e_3, e_4, e_5, e_6$ isomorphic to the elements of the Group C_{3v} with the correspondences

$$e_1 \rightarrow E,\ e_2 \rightarrow A,\ e_3 \rightarrow B,\ e_4 \rightarrow C,\ e_5 \rightarrow D,\ e_6 \rightarrow F$$

The multiplication table of these elements is shown in Table I.

TABLE I. MULTIPLICATION TABLE OF THE ALGEBRA

	e_1	e_2	e_3	e_4	e_5	e_6
e_1	e_1	e_2	e_3	e_4	e_5	e_6
e_2	e_2	e_1	e_5	e_6	e_3	e_4
e_3	e_3	e_6	e_1	e_5	e_4	e_2
e_4	e_4	e_5	e_6	e_1	e_2	e_3
e_5	e_5	e_4	e_2	e_3	e_6	e_1
e_6	e_6	e_3	e_4	e_2	e_1	e_5

We now list a few properties of this algebra

1. The algebra has a modulus e_1 because $e_1 e_i = e_i\, e_1 = e_i$ for $i = 1,2,..6$.
2. The algebra has an idempotent that is also a principal idempotent $e_1^2 = e_1$, there does not exist an e_j for which $e_1 e_j = 0$ or $e_j e_1 = 0$.
3. The subset of elements e_1, e_5, e_6 forms an algebra in itself; this is then a subalgebra of the algebra (also with a modulus). However, the products $e_i e_1$, $e_1 e_i$, $e_i e_5$, $e_5 e_i$, $e_i e_6$, $e_6 e_i$ where e_i is any element of the main algebra, are

not all members of this subalgebra. For instance, $e_4 e_5 = e_2$, and e_2 is not in the subset. When this does not happen, the subalgebra is not an "invariant subalgebra". An algebra that does not have an invariant subalgebra is called a "*simple*" *algebra*. This, then, is a simple algebra.

4. This algebra does not have a nilpotent element, that is, $e_i^n \neq 0$ for any e_i and for any n. It is obvious that the subalgebra does not have a nilpotent element either. Algebras having no nilpotent invariant subalgebras are called *semisimple*. Our algebra is, therefore, a semisimple algebra. Naturally, all simple algebras are semisimple.

5. The following theorems are at once satisfied:

   A simple algebra always contains an idempotent element.

   Every algebra that does not possess a modulus has a nilpotent invariant subalgebra.

   A semisimple algebra always has a modulus.

   If an algebra has a modulus e_1, this element is a principal idempotent and the only one.

The important consequence of the algebra being semisimple is that it is equivalent to a direct sum of total matrix algebras. This can be seen a little more explicitly with the help of the characteristic equation.

The characteristic equation of x is

$$\begin{vmatrix} x_1-\omega & x_2 & x_3 & x_4 & x_6 & x_5 \\ x_2 & (x_1-\omega) & x_5 & x_6 & x_4 & x_3 \\ x_3 & x_6 & (x_1-\omega) & x_5 & x_2 & x_4 \\ x_4 & x_5 & x_6 & (x_1-\omega) & x_3 & x_2 \\ x_5 & x_4 & x_2 & x_3 & (x_1-\omega) & x_6 \\ x_6 & x_3 & x_4 & x_2 & x_5 & (x_1-\omega) \end{vmatrix} = 0 \qquad (21)$$

a polynomial in ω of the sixth degree. A simple algebraic multiplication goes to show that there are three linear sets that commute with every element of the algebra:

$$\sigma_1 = e_1, \quad \sigma_2 = e_2 + e_3 + e_4, \quad \sigma_3 = e_5 + e_6$$

$$\sigma_i e_j - e_j \sigma_i \equiv [\sigma_i, e_j] = 0 \quad \text{for all } i, j \tag{22}$$

The σ_i themselves are elements of an Abelien algebra of order 3 with the multiplication table:

	σ_1	σ_2	σ_3
σ_1	σ_1	σ_2	σ_3
σ_2	σ_2	$2(\sigma_1 + \sigma_3)$	$2\sigma_2$
σ_3	σ_3	$2\sigma_2$	$2\sigma_1 + \sigma_3$

Since the algebra is semisimple, the characteristic equation reduces to a product of three factors, there being as many factors as there are commuting elements in the algebra, and these will be raised to powers 1, 1, and 2 respectively satisfying

$$\sum_\mu (n_\mu \times n_\mu) = (1 \times 1) + (1 \times 1) + (2 \times 2) = 6 \tag{23}$$

The first two factors are of first degree in x_i, and each occurs once in the product; the third factor will be of second degree and occur twice in the product. Each irreducible factor is thus raised to an exponent equal to its degree. The algebra is equal to a direct sum of two 1 x 1 matrix algebras and two 2 x 2 matrix algebras. Actual reduction of a characteristic determinant was shown in the earlier example, where each element was a commuting element in the algebra.

These results we owe to Cartan, Frobenius and Poincare.

CHAPTER TWO

GROUPS

The following Cayley multiplication table abstractly defines a finite group of six elements, a group of order 6. Let us denote the group by G and its order by g.

	E	A	B	C	D	F
E	E	A	B	C	D	F
A	A	E	D	F	B	C
B	B	F	E	D	C	A
C	C	D	F	E	A	B
D	D	C	A	B	F	E
F	F	B	C	A	E	D

These multiplications should be read as AB = A x B = D, BA = B x A = F, and so on. It is interesting to note that the elements can also be written in terms of two generators D and B. Thus

$$A = BD^2 \quad C = BD \quad F = D^2 \quad E = D^3 = (DB)^2 = (BD)^2 \quad (1)$$

we notice the multiplication is not commutative, and so the group is not Abelien.

There is more than one way of 'realizing' this group. For instance, the elements can be operations of permutation on three objects labeled, say, 1, 2, and 3. We associate the operation "1 is replaced by 2, 2 is replaced by 3, and 3 is replaced by 1" with the element D

$$\begin{pmatrix} 1 & 2 & 3 \\ 2 & 3 & 1 \end{pmatrix} = D$$

and the other elements with the remaining five permutations

$$\begin{pmatrix} 1 & 2 & 3 \\ 1 & 3 & 2 \end{pmatrix} = A \quad \begin{pmatrix} 1 & 2 & 3 \\ 3 & 2 & 1 \end{pmatrix} = B \quad \begin{pmatrix} 1 & 2 & 3 \\ 2 & 1 & 3 \end{pmatrix} = C \quad \begin{pmatrix} 1 & 2 & 3 \\ 3 & 1 & 2 \end{pmatrix} = F \quad \begin{pmatrix} 1 & 2 & 3 \\ 1 & 2 & 3 \end{pmatrix} = E \tag{2}$$

If we define AB as the operation A performed after the permutation corresponding to B, it is easy to see that the result is the same as directly performing operation D

$$\begin{pmatrix} 1 & 2 & 3 \\ 3 & 2 & 1 \end{pmatrix} \text{ followed by } \begin{pmatrix} 3 & 2 & 1 \\ 2 & 3 & 1 \end{pmatrix} \text{ is the same as } \begin{pmatrix} 1 & 2 & 3 \\ 2 & 3 & 1 \end{pmatrix}$$

The other multiplications in the group table are likewise realized. The number of elements in the permutation group or symmetric group on three objects, S_3, happens to be 6. Permutations on n objects form a group S_n, and its order is n! Permutations groups are of great importance in group theory because in Cayley theorem every finite group of order n is isomorphic with a subgroup of the symmetric group S_n (of order n!).

In Chapter 3 we will see another way of realizing this group by a set of geometrical symmetry operations of rotations and mirror reflections performed on a molecular aggregate of atoms. The corresponding operations are denoted by $E = E$, $A = \sigma_v$, $B = \sigma_{v''}$, $C = \sigma_{v'}$, $D = C_3^{\,2}$, $F = C_3$, and this group is called $G = C_{3v}$. A third and interesting way of realizing this group is by choosing six 2 x 2 matrices and defining multiplication as matrix multiplication:

$$\underset{E}{\begin{pmatrix} 1 & 0 \\ 0 & 1 \end{pmatrix}} \underset{A}{\begin{pmatrix} -1 & 0 \\ 0 & 1 \end{pmatrix}} \underset{B}{\begin{pmatrix} \frac{1}{2} & -\frac{\sqrt{3}}{2} \\ -\frac{\sqrt{3}}{2} & -\frac{1}{2} \end{pmatrix}} \underset{C}{\begin{pmatrix} \frac{1}{2} & \frac{\sqrt{3}}{2} \\ \frac{\sqrt{3}}{2} & -\frac{1}{2} \end{pmatrix}} \underset{D}{\begin{pmatrix} -\frac{1}{2} & \frac{\sqrt{3}}{2} \\ -\frac{\sqrt{3}}{2} & -\frac{1}{2} \end{pmatrix}} \underset{F}{\begin{pmatrix} -\frac{1}{2} & -\frac{\sqrt{3}}{2} \\ \frac{\sqrt{3}}{2} & -\frac{1}{2} \end{pmatrix}}$$

$$AB = \begin{pmatrix} -1 & 0 \\ 0 & 1 \end{pmatrix} \begin{pmatrix} \frac{1}{2} & -\frac{\sqrt{3}}{2} \\ -\frac{\sqrt{3}}{2} & -\frac{1}{2} \end{pmatrix} = \begin{pmatrix} -\frac{1}{2} & \frac{\sqrt{3}}{2} \\ -\frac{\sqrt{3}}{2} & -\frac{1}{2} \end{pmatrix} = D$$

$$BA = \begin{pmatrix} \frac{1}{2} & -\frac{\sqrt{3}}{2} \\ -\frac{\sqrt{3}}{2} & -\frac{1}{2} \end{pmatrix} \begin{pmatrix} -1 & 0 \\ 0 & 1 \end{pmatrix} = \begin{pmatrix} -\frac{1}{2} & -\frac{\sqrt{3}}{2} \\ \frac{\sqrt{3}}{2} & -\frac{1}{2} \end{pmatrix} = F \tag{3}$$

We notice that the elements of the permutation group S_3 can be put in a one-to-one correspondence with the elements of the molecular symmetry group C_{3v} as with the elements of the group of matrices. This one-to-one correspondence that is unique and reciprocal, which is preserved in multiplication, is called *isomorphism.* On the other hand, the multiplication table can also be realized by the following set of numbers, multiplication here being ordinary multiplication:

E	A	B	C	D	F
1	-1	-1	-1	1	1

Here, however, the correspondence is not one to one but many to one, because the number 1 stands for the elements E, D, and F at the same time. Such a many to one correspondence is called *homomorphism;* thus the group S_3 is isomorphic to C_{3v}, but homomorphic to the group of the above numbers.

SUBGROUPS, COSETS AND CLASSES

Out of the set of six elements of G we can choose a subset of three elements (including the identity) E, D, F which themselves satisfy all the group postulates, with the following multiplication table:

	E	D	F
E	E	D	F
D	D	F	E
F	F	E	D

This is called a *subgroup* (say H) of order h=3 of the main group G of order g=6. There are three other subgroups, each of order 2: E,A; E,B; E,C. We also can have the rather trivial subgroup of order 1, consisting of the identity E alone; this is not called a proper subgroup. Thus the group of order 6 has proper subgroups of orders 2 and 3. This follows Lagrange's theorem in finite group theory, which says that the order of a subgroup is a whole number divisor of the order of the whole group. Thus, for instance, a group of order 9 cannot have a subgroup of order 2 or 4 or 6. Notice the identity has to be one of the elements of any subgroup. The ratio of the order of the group to that of the subgroup g/h is called the index of the subgroup. The index of the subgroup above is 2.

A, B and C are not members of the subgroup H but certainly members of G. From the multiplication table we readily get

$$\begin{array}{lll} EA = A & EB = B & EC = C \\ DA = G & DB = A & DC = B \\ FA = B & FB = C & FC = A \end{array} \qquad (4)$$

The set EX, DX, FX, where X is an element of the group G but not an element of its subgroup H, is called the *right coset* corresponding to H. Between the elements of H and the elements of the right coset all the six elements of the group G are contained. Symbolically,

$$G = H + HA$$

In a similar way a *left coset* is defined, and the group can again be made up of the subgroup and the left cosets

$$G = H + AH$$

The coset can never be a subgroup, because the identity element is not a member of this set.

From the multiplication table we have the following relations:

$$D^{-1} = F \qquad\qquad D\,A\,D^{-1} = D\,A\,F = D\,C = B$$

If we can find an element (D) such that for a given element (A) we can get another element of the group by the operation $D\,A\,D^{-1} = B$, then A and B are said to be *conjugate* to each other, and the process DAD^{-1} is called *conjugation* of A with D, resembling a similarity transformation in matrix theory. If we now choose any one of the elements A, B, C and perform its conjugation with all the elements of the group, the result will be either A or B or C (in some order). Thus, A, B and C are a set of *self-conjugate* elements and such a set is called a *class*, (say, K_2). Similarly, D and F form another class (say, K_3). It is obvious that the identity element E, which commutes with all the elements of the group, is a class by itself (say, K_1). Thus a group G can be uniquely divided into classes; K_1, K_2, and K_3 in our case. Notice, on the other hand, the set A, C, D does not form a class:

$$B\,C\,B^{-1} = B\,C\,B = B\,F = A$$

$$B\,A\,B^{-1} = B\,A\,B = B\,D = C$$

but

$$B\,D\,B^{-1} = B\,D\,B = B\,A = F \qquad (5)$$

and F is not a member of our set, although A and C are. A class can, therefore, be determined by knowing one of its elements and then conjugating this with all the members of the group:

$$EAE^{-1} = A, \quad AAA^{-1} = A, \quad BAB^{-1} = C, \quad CAC^{-1} = B,$$

$$DAD^{-1} = B, \quad FAF^{-1} = C$$

Notice in the group of matrices all the elements in a given class have the same trace, the traces for K_1, K_2, and K_3 being, respectively, 2, 0, and -1. It is easy to see that each element of an Abelian group forms a class by itself.

A subgroup is called an *invariant* (or normal) subgroup if all its elements consist of entire classes. In the elements of H, E belongs to the class K_1 and D, F to the class K_3, and there are no elements besides these two classes. H is thus an invariant subgroup of G. Groups that have no invariant subgroups are called *simple* groups; if a group has a non-Abelian invariant subgroup, it is called *semisimple*. Since H is non-Abelian, though invariant, G should be classified as a *semisimple* group. Semisimple groups are of importance in the theory of continuous Lie groups.

It is easy to see from the multiplication table that for the invariant subgroup $\underset{\sim}{H}$ the left and right cosets are the same. Furthermore, H and $h \equiv AH$, considered as collectively two elements, form a group with the following multiplication table:

	H	h
H	H	h
h	h	H

For instance, H x h = (E,D,F) x (A,B,C) = (A,B,C) = h. The invariant subgroup plays the role of the identity element. This is called a Factor Group and its order (2) is the index of H. This factor group is isomorphic with the symmetry group C_2.

INNER AUTOMORPHISMS AND SEMI-DIRECT PRODUCTS

An isomorphism of a group with itself, that is, a one-to-one correspondence between elements of the group preserving multiplication, is called an *automorphism*. If this correspondence is brought about by conjugation, it is called an *inner automorphism*. In the group G let

$$\left.\begin{array}{l} E \quad E \\ A \leftrightarrow C \\ B \leftrightarrow B \\ C \leftrightarrow A \\ D \leftrightarrow F \\ F \leftrightarrow D \end{array}\right\}$$

such that, for example,

$$\begin{array}{ccc} A\ B & = & D \\ \downarrow\ \downarrow & & \downarrow \\ C\ B & = & F \end{array}$$

since D is mapped into F the multiplication is preserved. (6)

This, then, is an automorphism. But it is easy to see that if we conjugate all the elements of the group with the fixed element B, we get

$$BEB^{-1} = E,\ BAB^{-1} = C,\ BBB^{-1} = B,\ BCB^{-1} = A,$$
$$BDB^{-1} = F,\ BFB^{-1} = D \qquad (7)$$

The above isomorphism is, then, an inner automorphism <u>induced by B</u>. Formally, this is written

$$B\ G\ B^{-1} = G \qquad B \text{ fixed}$$

The other four inner automorphisms are given below. The set E A B C D F, which incidentally is an inner automorphism induced by the identity, is mapped into

E	B	A	C	F	D	(induced by C)	
E	A	C	B	F	D	(induced by A)	
E	B	C	A	D	F	(induced by D)	
E	C	A	B	D	F	(induced by F)	(8)

We see that in all inner automorphisms the elements E, D, and F are mapped into themselves. Thus an *invariant subgroup remains invariant under all inner automorphisms of the group.*

Consider the Abelian cyclic group of order 3 ($C=C_3$) defined by the multiplication table

	E	A_1	A_2
E	E	A_1	A_2
A_1	A_1	A_2	E
A_2	A_2	E	A_1

In the symmetry group C_3, A_1 and A_2 correspond to rotations through angle $2\pi/3$ and $4\pi/3$, respectively. It is easy to prove that the following one-to-one correspondence preserve multiplication and hence are automorphisms of G_1 ($=C_3$):

$$E = \alpha \to \left[\begin{matrix} E & A_1 & A_2 \\ \downarrow & \downarrow & \downarrow \\ E & A_1 & A_2 \end{matrix}\right.$$

$$\beta \to \left[\begin{matrix} E & A_1 & A_2 \\ \downarrow & \downarrow & \downarrow \\ E & A_2 & A_1 \end{matrix}\right.$$

Furthermore, α and β are elements of a group of automorphisms with the multiplication table

	α(=B)	β
α	E	β
β	β	E

where by multiplication we mean, for instance, the mapping β

followed by the mapping $\alpha = \alpha\beta$. From the elements of the group G_1 and the elements of the *group of its automorphisms* G_a we form ordered pairs as new elements:

$$
\begin{array}{llll}
(E,\alpha) = (E,E) \equiv E & & (A_1,\beta) \equiv B & \\
(E,\beta) \equiv A & & (A_2,\alpha) = (A_2,E) \equiv D & \\
(A_1,\alpha) \equiv (A_1,E) \equiv F & & (A_2,\beta) \equiv C & \quad (9)
\end{array}
$$

These six ordered pairs form a group called the *semidirect product* of G_1 and G_a written $G_1 \boxed{s} G_a$. The rule of combination or multiplication in this semidirect product group is the following:

$$B\ D \equiv (A_1,\beta)\ (A_2,\alpha) = (A_1,\beta(A_2),\ \beta\alpha) \qquad \{=C\}$$

where $\beta(A_2)$ is the image of A_2 under the automorphism β. Now, $\beta(A_2) = A_1$, and we thus have $(A_1A_1,\ \beta\alpha) = (A_2,\beta) \equiv C$, since $A_1^{\ 2} = A_2$ according to the multiplication table of group G_1 and $\beta\alpha = \beta$ according to the automorphism group G_a. Another multiplication, by way of illustration, is the following:

$$D\ B \equiv (A_2,\alpha)\ (A_1,\beta) = (A_2\ \alpha(A_1),\ \alpha\beta) = (A_2A_1,\ \alpha\beta) = (E\ \beta) = A$$

If we proceed in this way, it is easily seen that the semidirect product is just the group G isomorphic to S_3. Formally, we write

$$G_1 \boxed{s} G_a = G$$

CHAPTER THREE

REPRESENTATIONS AND CHARACTERS

REDUCIBLE AND IRREDUCIBLE REPRESENTATIONS

If the following 3 x 3 matrix corresponds to the element E in the symmetric group S_3

$$D(E) = \begin{pmatrix} 1 & 2 & 3 \\ 1 & 2 & 3 \end{pmatrix} \rightarrow \begin{pmatrix} 1 & 0 & 0 \\ 0 & 1 & 0 \\ 0 & 0 & 1 \end{pmatrix}$$

it is obvious that the element C is associated with

$$D(C) = \begin{pmatrix} 0 & 1 & 0 \\ 1 & 0 & 0 \\ 0 & 0 & 1 \end{pmatrix}$$

and similarly, the other correspondences are

$$D(A) = \begin{pmatrix} 1 & 0 & 0 \\ 0 & 0 & 1 \\ 0 & 1 & 0 \end{pmatrix}; \; D(B) = \begin{pmatrix} 0 & 0 & 1 \\ 0 & 1 & 0 \\ 1 & 0 & 0 \end{pmatrix}; \; D(D) = \begin{pmatrix} 0 & 0 & 1 \\ 1 & 0 & 0 \\ 0 & 1 & 0 \end{pmatrix};$$

$$D(F) = \begin{pmatrix} 0 & 1 & 0 \\ 0 & 0 & 1 \\ 1 & 0 & 0 \end{pmatrix} \qquad (1)$$

A simple calculation shows that these elements satisfy the Cayley multiplication table, multiplication of two elements now meaning matrix multiplication. This set of matrices is called a three-dimensional representation of the group S_3. By means of a non-singular matrix S we can make a similarity transformation and get another set of matrices D'(E); D'(A)...., which will also satisfy the same multiplication table. Two such sets are said to be *equivalent representations*. In particular, the matrix

$$S = \begin{pmatrix} \frac{1}{\sqrt{3}} & \frac{1}{\sqrt{3}} & \frac{1}{\sqrt{3}} \\ 0 & -\frac{1}{\sqrt{2}} & \frac{1}{\sqrt{2}} \\ \frac{2}{\sqrt{3}} & -\frac{1}{\sqrt{6}} & -\frac{1}{\sqrt{6}} \end{pmatrix}; \quad S^{-1} = \begin{pmatrix} \frac{1}{\sqrt{3}} & 0 & \sqrt{\frac{2}{3}} \\ \frac{1}{\sqrt{3}} & -\frac{1}{\sqrt{2}} & -\frac{1}{\sqrt{6}} \\ \frac{1}{\sqrt{3}} & \frac{1}{\sqrt{2}} & -\frac{1}{\sqrt{6}} \end{pmatrix}$$

reduces *every* matrix through the similarity transformation into a *direct sum* of two lower-dimensional matrices--a two-dimensional matrix and a one-dimensional matrix

$$S\ D(A)\ S^{-1} = \begin{pmatrix} \frac{1}{\sqrt{3}} & \frac{1}{\sqrt{3}} & \frac{1}{\sqrt{3}} \\ 0 & -\frac{1}{\sqrt{2}} & \frac{1}{\sqrt{2}} \\ \sqrt{\frac{2}{3}} & -\frac{1}{\sqrt{6}} & -\frac{1}{\sqrt{6}} \end{pmatrix} \begin{pmatrix} 1 & 0 & 0 \\ 0 & 0 & 1 \\ 0 & 1 & 0 \end{pmatrix} \begin{pmatrix} \frac{1}{\sqrt{3}} & 0 & \sqrt{\frac{2}{3}} \\ \frac{1}{\sqrt{3}} & -\frac{1}{\sqrt{2}} & -\frac{1}{\sqrt{6}} \\ \frac{1}{\sqrt{3}} & \frac{1}{\sqrt{2}} & \frac{1}{\sqrt{6}} \end{pmatrix}$$

$$= \begin{pmatrix} 1 & 0 & 0 \\ 0 & -1 & 0 \\ 0 & 0 & 1 \end{pmatrix} = (1) \oplus \begin{pmatrix} -1 & 0 \\ 0 & 1 \end{pmatrix}$$

$$S\ D(B)\ S^{-1} = \begin{pmatrix} 1 & 0 & 0 \\ 0 & \frac{1}{2} & \frac{\sqrt{3}}{2} \\ 0 & \frac{\sqrt{3}}{2} & -\frac{1}{2} \end{pmatrix} = (1) \oplus \begin{pmatrix} \frac{1}{2} & \frac{\sqrt{3}}{2} \\ \frac{\sqrt{3}}{2} & -\frac{1}{2} \end{pmatrix}$$

(2)

$$S\,D(C)\,S^{-1} = \begin{pmatrix} 1 & 0 & 0 \\ 0 & \frac{1}{2} & -\frac{\sqrt{3}}{2} \\ 0 & -\frac{\sqrt{3}}{2} & -\frac{1}{2} \end{pmatrix}; \quad S\,D(D)\,S^{-1} = \begin{pmatrix} 1 & 0 & 0 \\ 0 & -\frac{1}{2} & -\frac{\sqrt{3}}{2} \\ 0 & \frac{\sqrt{3}}{2} & -\frac{1}{2} \end{pmatrix}$$

$$S\,D(F)\,S^{-1} = \begin{pmatrix} 1 & 0 & 0 \\ 0 & -\frac{1}{2} & \frac{\sqrt{3}}{2} \\ 0 & -\frac{\sqrt{3}}{2} & -\frac{1}{2} \end{pmatrix} = (1) \oplus \begin{pmatrix} -\frac{1}{2} & \frac{\sqrt{3}}{2} \\ -\frac{\sqrt{3}}{2} & -\frac{1}{2} \end{pmatrix} \tag{3}$$

The representation D(C), D(A)....is for this reason called a *reducible representation*. It can be shown that the 2 x 2 matrices cannot be further reduced through any similarity transformation in like manner, and hence this latter set is called an *irreducible representation*. In other words, irreducible representations are those matrices that cannot, by a single similarity transformation, be expressed as direct sums of lower-dimensional matrices. Since an equivalent irreducible representation can always be generated from a given irreducible representation, we assume henceforth that the irreducible representations consist of sets of *unitary* matrices. The trace of the irreducible representation, which is invariant to a similarity transformation, is called a *character* and the trace of the reducible representation is called a *compound character*.' Because of Cayley's theorem we can find the irreducible representations and characters of any finite group if we have knowledge of these for an appropriate symmetric group.

We take the symmetric group S_4, the group of permutations on four objects, as an example to illustrate important theorems on representation and characters. The finite group $G(=S_4)$ is of order $g = 4! = 24$. The elements of the group are given below. Its Cayley multiplication table is given in Table I.

$$\underset{E}{\begin{pmatrix} 1 & 2 & 3 & 4 \\ 1 & 2 & 3 & 4 \end{pmatrix}} \quad \underset{A}{\begin{pmatrix} 1 & 2 & 3 & 4 \\ 1 & 4 & 2 & 3 \end{pmatrix}} \quad \underset{B}{(1\ 3\ 4\ 2)} \quad \underset{C}{(1\ 2\ 4\ 3)}$$

TABLE I. CAYLEY MULTIPLICATION TABLE FOR S_4

	E	A	B	C	D	F	G	H	I	J	K	L	M	N	O	P	Q	R	S	T	U	V	W	X	
E	E	A	B	C	D	F	G	H	I	J	K	L	M	N	O	P	Q	R	S	T	U	V	W	X	E
A	A	B	E	D	F	C	R	U	O	K	N	I	V	J	L	H	T	X	Q	S	P	W	M	G	A
B	B	E	A	F	C	D	X	P	L	N	J	O	W	K	I	U	S	G	T	Q	H	M	V	R	B
C	C	F	D	E	B	A	K	O	U	R	G	P	T	X	H	L	V	J	W	M	I	Q	S	N	C
D	D	C	F	A	E	B	N	L	P	X	R	H	S	G	U	I	W	K	M	V	O	T	Q	J	D
F	F	D	C	B	A	E	J	I	H	G	X	U	Q	R	P	O	M	N	V	W	L	S	T	K	F
G	G	U	S	K	W	I	E	J	F	H	C	T	P	Q	R	M	N	C	B	L	A	X	D	V	G
H	H	T	X	V	L	J	I	E	G	F	S	D	R	P	Q	N	O	M	K	A	W	C	U	B	H
I	I	W	K	S	U	G	H	F	J	E	V	A	N	O	M	R	P	Q	X	D	T	B	L	C	I
J	J	L	V	X	T	H	F	G	E	I	B	W	O	M	N	Q	R	P	C	U	D	K	A	S	J
K	K	I	W	G	S	U	C	R	A	O	E	M	L	V	J	T	X	H	D	P	F	N	B	Q	K
L	L	V	J	T	H	X	P	D	N	B	M	E	K	I	W	G	U	S	R	C	Q	A	O	F	L
M	M	N	O	P	R	Q	T	S	V	W	L	K	E	A	B	C	F	D	H	G	X	I	J	U	M
N	N	O	M	R	Q	P	D	X	B	L	A	V	I	W	K	S	G	U	F	H	C	J	E	T	N
O	O	M	N	Q	P	R	U	C	K	A	W	B	J	L	V	X	H	T	G	F	S	E	I	D	O
P	P	Q	R	M	O	N	L	B	X	D	T	C	G	U	S	E	I	W	J	E	V	F	H	A	P
Q	Q	R	P	O	N	M	W	V	S	T	U	X	F	D	C	B	E	A	I	J	K	H	G	L	Q
R	R	P	Q	N	M	O	A	K	C	U	D	S	H	T	X	V	J	L	E	I	B	G	F	W	R
S	S	G	U	I	K	W	V	M	T	Q	H	R	D	C	F	A	B	E	L	N	J	P	X	O	S
T	T	X	H	L	J	V	M	W	Q	S	P	G	C	F	D	E	A	B	O	K	N	U	R	I	T
U	U	S	G	W	I	K	O	A	R	C	Q	F	X	H	T	J	L	V	N	B	M	D	P	E	U
V	V	J	L	H	X	T	S	Q	W	M	I	N	A	B	E	D	C	F	U	R	G	O	K	P	V
W	W	K	I	U	G	S	Q	T	M	V	O	J	B	E	A	F	D	O	P	X	R	L	N	H	W
X	X	H	T	J	V	L	B	N	D	P	F	Q	U	S	G	W	K	I	A	O	E	R	C	M	X
	E	A	B	C	D	F	G	H	I	J	K	L	M	N	O	P	Q	R	S	T	U	V	W	X	

D	F	G	H
(1 4 3 2)	(1 3 2 4)	(2 1 3 4)	(3 2 1 4)
I	J	K	L
(2 3 1 4)	(3 1 2 4)	(2 1 4 3)	(3 4 1 2)
M	N	O	P
(4 3 2 1)	(4 1 3 2)	(4 2 1 3)	(4 3 1 2)
Q	R	S	T
(4 2 3 1)	(4 1 2 3)	(2 3 4 1)	(3 4 2 1)
U	V	W	X
(2 4 1 3)	(3 2 4 1)	(2 4 3 1)	(3 1 4 2)

(4)

The classes of S_4 are equal to the number of distinct partitions of 4

$$K_1 : (E) \rightarrow 1 + 1 + 1 + 1 \quad (1^4) \quad g_1 = 1$$

$$K_2 : (C,D,F,G,H,Q) \rightarrow 1 + 1 + 2 \quad (1^2 2) \quad g_2 = 6$$

$$K_3 : (K,L,M) \rightarrow 2 + 2 \quad (2^2) \quad g_3 = 3$$

$$K_4 : (A,B,I,J,N,O,V,W) \rightarrow 1 + 3 \quad (31) \quad g_4 = 8$$

$$K_5 : (P,R,S,T,U,X) \rightarrow 4 + 0 \quad (4) \quad g_5 = 6 \qquad (5)$$

g_i is the number of elements in class K_i. The partitions can be elegantly shown in the form of tableaux developed in a series of papers published in the Proceedings of the London Mathematical Society at the turn of the Century by Rev. A. Young.

K_5: (4) → [] [] [] []

K_1 : (1^4)

[]
[]
[]
[]

K_4 : (31) [] [] []
[]

K_3 : (2^2) [] []
[] []

K_2 : (21^2) [] []
[]
[] (6)

The arrangement of the boxes corresponds to the partitions in an obvious way. According to representation theory

1. The number of inequivalent irreducible representations is equal to the number of classes,
2. The sum of squares of the dimensions of the irreducible representations equals the order of the group

$$n_1^2 + n_2^2 + n_3^2 + n_4^2 + n_5^2 = g,$$

n_μ being the dimension of the μ-th irreducible representation,

3. In an irreducible representation ν all the elements of a class i have the same character $\chi_i^{(\nu)}$.

Inserting numbers in the boxes with the convention that the integers appear in increasing order, horizontally as well as vertically, we define a *standard tableau:*

[1] [2] [3]
[4]

is a standard tableau, but

[1] [3] [2]
[4] (7)

is not. There is, however, more than one way of placing numbers without violating the "standard" prescription. The number of ways in which this can be done, or the number of standard tableaux, is equal to the dimension of the representation. We have, for instance,

[1] [2] [3]   [1] [2] [4]   [1] [3] [4]
[4]   [3]   [2] (8)

and this corresponds to a three-dimensional representation. If we proceed this way, we can easily establish that S_4 has 2 one-dimensional, 1 2-dimensional, and 2 three-dimensional irreducible representations, as shown in the following.

Irreducible Representations of S_4 Isomorphic to O and T_d

$D^{(5)}$: All elements are represented by the number 1 (one-dimensional matrix)

$D^{(1)}$: E is represented by 1 as also A,B,I,J,K,L,M,N,O,V,W and -1 represents C,D,F,G,H,P,Q,R,S,T,U,X.

$D^{(3)}$: E,K,L,M are represented by $\begin{pmatrix} 1 & 0 \\ 0 & 1 \end{pmatrix}$, G,C,T,P are represented by $\begin{pmatrix} 1 & 0 \\ 0 & -1 \end{pmatrix}$, F,Q,U,X are represented by

$$\begin{pmatrix} -\frac{1}{2} & \frac{\sqrt{3}}{2} \\ \frac{\sqrt{3}}{2} & \frac{1}{2} \end{pmatrix}$$, D,H,R,S are represented by

$$\begin{pmatrix} -\frac{1}{2} & -\frac{\sqrt{3}}{2} \\ -\frac{\sqrt{3}}{2} & \frac{1}{2} \end{pmatrix}$$, A,I,N,V are represented by

$$\begin{pmatrix} -\frac{1}{2} & \frac{\sqrt{3}}{2} \\ -\frac{\sqrt{3}}{2} & -\frac{1}{2} \end{pmatrix}$$ and B,J,O,W are represented by

$$\begin{pmatrix} -\frac{1}{2} & -\frac{\sqrt{3}}{2} \\ \frac{\sqrt{3}}{2} & -\frac{1}{2} \end{pmatrix}$$

$\underline{D^{(2)}}: \to$

$$D(A) = \begin{pmatrix} -\frac{i}{2} & \frac{i}{\sqrt{2}} & -\frac{i}{2} \\ \frac{1}{\sqrt{2}} & 0 & -\frac{1}{\sqrt{2}} \\ \frac{i}{2} & \frac{i}{\sqrt{2}} & \frac{i}{2} \end{pmatrix}$$

$$x = 0$$

$$D(B) = \begin{pmatrix} \frac{i}{2} & \frac{1}{\sqrt{2}} & \frac{i}{2} \\ -\frac{i}{\sqrt{2}} & 0 & -\frac{i}{\sqrt{2}} \\ \frac{i}{2} & -\frac{1}{\sqrt{2}} & -\frac{i}{2} \end{pmatrix}$$

$$x = 0$$

$$D(C) = \begin{pmatrix} 0 & 0 & i \\ 0 & -1 & 0 \\ -i & 0 & 0 \end{pmatrix}$$

$$x - -1$$

$$D(D) = \begin{pmatrix} -\frac{1}{2} & -\frac{i}{\sqrt{2}} & \frac{1}{2} \\ \frac{i}{\sqrt{2}} & 0 & \frac{i}{\sqrt{2}} \\ \frac{1}{2} & -\frac{i}{\sqrt{2}} & -\frac{1}{2} \end{pmatrix}$$

$$x = -1$$

$$D(F) = \begin{pmatrix} -\frac{1}{2} & -\frac{1}{\sqrt{2}} & -\frac{1}{2} \\ -\frac{1}{\sqrt{2}} & 0 & \frac{1}{\sqrt{2}} \\ -\frac{1}{2} & \frac{1}{\sqrt{2}} & -\frac{1}{2} \end{pmatrix}$$

$$\chi = -1$$

$$D(G) = \begin{pmatrix} 0 & 0 & -i \\ 0 & -1 & 0 \\ i & 0 & 0 \end{pmatrix}$$

$$\chi = -1$$

$$D(H) = \begin{pmatrix} -\frac{1}{2} & \frac{i}{\sqrt{2}} & \frac{1}{2} \\ -\frac{i}{\sqrt{2}} & 0 & -\frac{i}{\sqrt{2}} \\ \frac{1}{2} & \frac{i}{\sqrt{2}} & -\frac{1}{2} \end{pmatrix}$$

$$\chi = -1$$

$$D(I) = \begin{pmatrix} \frac{i}{2} & -\frac{i}{\sqrt{2}} & \frac{i}{2} \\ \frac{1}{\sqrt{2}} & 0 & -\frac{1}{\sqrt{2}} \\ -\frac{i}{2} & -\frac{i}{\sqrt{2}} & -\frac{i}{2} \end{pmatrix}$$

$$\chi = 0$$

$$D(J) = \begin{pmatrix} -\frac{i}{2} & \frac{1}{\sqrt{2}} & \frac{i}{2} \\ \frac{i}{\sqrt{2}} & 0 & \frac{i}{\sqrt{2}} \\ -\frac{i}{2} & -\frac{1}{\sqrt{2}} & \frac{i}{2} \end{pmatrix}$$

$$\chi = 0$$

$$D(K) = \begin{pmatrix} -1 & 0 & 0 \\ 0 & 1 & 0 \\ 0 & 0 & -1 \end{pmatrix}$$

$$\chi = -1$$

$$D(L) = \begin{pmatrix} 0 & 0 & -1 \\ 0 & -1 & 0 \\ -1 & 0 & 0 \end{pmatrix}$$

$$\chi = -1$$

$$D(M) \quad \begin{pmatrix} 0 & 0 & 1 \\ 0 & -1 & 0 \\ 1 & 0 & 0 \end{pmatrix}$$

$$\chi = -1$$

$$D(N) = \begin{pmatrix} \frac{i}{2} & \frac{i}{\sqrt{2}} & \frac{i}{2} \\ -\frac{1}{\sqrt{2}} & 0 & \frac{1}{\sqrt{2}} \\ -\frac{i}{2} & \frac{i}{\sqrt{2}} & -\frac{i}{2} \end{pmatrix}$$

$$\chi = 0$$

$$D(O) = \begin{pmatrix} \frac{i}{2} & -\frac{1}{\sqrt{2}} & -\frac{i}{2} \\ \frac{i}{\sqrt{2}} & 0 & \frac{i}{\sqrt{2}} \\ \frac{i}{2} & \frac{1}{\sqrt{2}} & -\frac{i}{2} \end{pmatrix}$$

$$\chi = 0$$

$$D(P) = \begin{pmatrix} -i & 0 & 0 \\ 0 & 1 & 0 \\ 0 & 0 & i \end{pmatrix}$$

$$\chi = 1$$

$$D(Q) = \begin{pmatrix} -\frac{1}{2} & +\frac{1}{\sqrt{2}} & -\frac{1}{2} \\ \frac{1}{\sqrt{2}} & 0 & -\frac{1}{\sqrt{2}} \\ -\frac{1}{2} & -\frac{1}{\sqrt{2}} & -\frac{1}{2} \end{pmatrix}$$

$$\chi = -1$$

$$D(R) = \begin{pmatrix} \frac{1}{2} & -\frac{i}{\sqrt{2}} & -\frac{1}{2} \\ -\frac{i}{\sqrt{2}} & 0 & -\frac{i}{\sqrt{2}} \\ -\frac{1}{2} & -\frac{i}{\sqrt{2}} & \frac{1}{2} \end{pmatrix}$$

$$\chi = 1$$

$$D(S) = \begin{pmatrix} \frac{1}{2} & \frac{i}{\sqrt{2}} & -\frac{1}{2} \\ \frac{i}{\sqrt{2}} & 0 & \frac{i}{\sqrt{2}} \\ -\frac{1}{2} & \frac{i}{\sqrt{2}} & \frac{1}{2} \end{pmatrix}$$

$$\chi = 1$$

$$D(T) = \begin{pmatrix} i & 0 & 0 \\ 0 & 1 & 0 \\ 0 & 0 & -i \end{pmatrix}$$

$$\chi = 1$$

$$D(U) = \begin{pmatrix} \frac{1}{2} & \frac{1}{\sqrt{2}} & \frac{1}{2} \\ -\frac{1}{\sqrt{2}} & 0 & \frac{1}{\sqrt{2}} \\ \frac{1}{2} & -\frac{1}{\sqrt{2}} & \frac{1}{2} \end{pmatrix}$$

$$\chi = 1$$

$$D(V) = \begin{pmatrix} -\frac{i}{2} & -\frac{i}{\sqrt{2}} & -\frac{i}{2} \\ -\frac{1}{\sqrt{2}} & 0 & \frac{1}{\sqrt{2}} \\ \frac{i}{2} & -\frac{i}{\sqrt{2}} & \frac{i}{2} \end{pmatrix}$$

$$\chi = 0$$

$$D(W) = \begin{pmatrix} -\frac{i}{2} & -\frac{1}{\sqrt{2}} & \frac{i}{2} \\ -\frac{i}{\sqrt{2}} & 0 & -\frac{i}{\sqrt{2}} \\ -\frac{i}{2} & \frac{1}{\sqrt{2}} & \frac{i}{2} \end{pmatrix}$$

$$\chi = 0$$

$$D(X) = \begin{pmatrix} \frac{1}{2} & -\frac{1}{\sqrt{2}} & \frac{1}{2} \\ \frac{1}{\sqrt{2}} & 0 & -\frac{1}{\sqrt{2}} \\ \frac{1}{2} & \frac{1}{\sqrt{2}} & \frac{1}{2} \end{pmatrix}$$

$$\chi = 1$$

$\underline{D^{(4)}}$: $\rightarrow$

$$D(A) = \begin{pmatrix} -\frac{i}{2} & \frac{i}{\sqrt{2}} & -\frac{i}{2} \\ \frac{1}{\sqrt{2}} & 0 & -\frac{1}{\sqrt{2}} \\ \frac{i}{2} & \frac{i}{\sqrt{2}} & \frac{i}{2} \end{pmatrix}$$

$$\chi = 0$$

$$D(B) = \begin{pmatrix} \frac{i}{2} & \frac{1}{\sqrt{2}} & -\frac{i}{2} \\ -\frac{i}{\sqrt{2}} & 0 & -\frac{i}{\sqrt{2}} \\ \frac{i}{2} & -\frac{1}{\sqrt{2}} & -\frac{i}{2} \end{pmatrix}$$

$$\chi = 0$$

$$D(C) = \begin{pmatrix} 0 & 0 & -i \\ 0 & 1 & 0 \\ i & 0 & 0 \end{pmatrix}$$

$$\chi = 1$$

$$D(D) = \begin{pmatrix} \frac{1}{2} & \frac{i}{\sqrt{2}} & -\frac{1}{2} \\ -\frac{i}{\sqrt{2}} & 0 & -\frac{i}{\sqrt{2}} \\ -\frac{1}{2} & \frac{i}{\sqrt{2}} & \frac{1}{2} \end{pmatrix}$$

$$\chi = 1$$

$$D(F) = \begin{pmatrix} \frac{1}{2} & \frac{1}{\sqrt{2}} & \frac{1}{2} \\ \frac{1}{\sqrt{2}} & 0 & -\frac{1}{\sqrt{2}} \\ \frac{1}{2} & -\frac{1}{\sqrt{2}} & \frac{1}{2} \end{pmatrix}$$

$$\chi = 1$$

$$D(G) = \begin{pmatrix} 0 & 0 & i \\ 0 & 1 & 0 \\ -i & 0 & 0 \end{pmatrix}$$

$$\chi = 1$$

$$D(H) = \begin{pmatrix} \frac{1}{2} & -\frac{i}{\sqrt{2}} & -\frac{1}{2} \\ \frac{i}{\sqrt{2}} & 0 & \frac{i}{\sqrt{2}} \\ -\frac{1}{2} & -\frac{i}{\sqrt{2}} & \frac{1}{2} \end{pmatrix}$$

$$\chi = 1$$

$$D(I) = \begin{pmatrix} \frac{i}{2} & -\frac{i}{\sqrt{2}} & \frac{i}{2} \\ \frac{1}{\sqrt{2}} & 0 & -\frac{1}{\sqrt{2}} \\ -\frac{i}{2} & -\frac{i}{\sqrt{2}} & -\frac{i}{2} \end{pmatrix}$$

$$\chi = 0$$

$$D(J) = \begin{pmatrix} -\frac{i}{2} & \frac{1}{\sqrt{2}} & \frac{i}{2} \\ \frac{i}{\sqrt{2}} & 0 & \frac{i}{\sqrt{2}} \\ -\frac{i}{2} & -\frac{1}{\sqrt{2}} & \frac{i}{2} \end{pmatrix}$$

$$\chi = 0$$

$$D(K) = \begin{pmatrix} -1 & 0 & 0 \\ 0 & 1 & 0 \\ 0 & 0 & -1 \end{pmatrix}$$

$$x = -1$$

$$D(L) = \begin{pmatrix} 0 & 0 & -1 \\ 0 & -1 & 0 \\ -1 & 0 & 0 \end{pmatrix}$$

$$x = -1$$

$$D(M) = \begin{pmatrix} 0 & 0 & 1 \\ 0 & -1 & 0 \\ 1 & 0 & 0 \end{pmatrix}$$

$$x = -1$$

$$D(N) \quad \begin{pmatrix} +\frac{i}{2} & \frac{i}{\sqrt{2}} & \frac{i}{2} \\ -\frac{1}{\sqrt{2}} & 0 & \frac{1}{\sqrt{2}} \\ -\frac{i}{2} & \frac{i}{\sqrt{2}} & -\frac{i}{2} \end{pmatrix}$$

$$x = 0$$

$$D(O) = \begin{pmatrix} \frac{i}{2} & -\frac{1}{\sqrt{2}} & -\frac{i}{2} \\ \frac{i}{\sqrt{2}} & 0 & \frac{i}{\sqrt{2}} \\ \frac{i}{2} & \frac{1}{\sqrt{2}} & -\frac{i}{2} \end{pmatrix}$$

$$x = 0$$

$$D(P) = \begin{pmatrix} i & 0 & 0 \\ 0 & -1 & 0 \\ 0 & 0 & -i \end{pmatrix}$$

$$x = -1$$

$$D(Q) = \begin{pmatrix} \frac{1}{2} & -\frac{1}{\sqrt{2}} & \frac{1}{2} \\ -\frac{1}{\sqrt{2}} & 0 & \frac{1}{\sqrt{2}} \\ \frac{1}{2} & \frac{1}{\sqrt{2}} & \frac{1}{2} \end{pmatrix}$$

$$x = 1$$

$$D(R) = \begin{pmatrix} -\frac{1}{2} & \frac{i}{\sqrt{2}} & \frac{1}{2} \\ \frac{i}{\sqrt{2}} & 0 & \frac{i}{\sqrt{2}} \\ \frac{1}{2} & \frac{i}{\sqrt{2}} & -\frac{1}{2} \end{pmatrix}$$

$$x = -1$$

$$D(S) = \begin{pmatrix} -\frac{1}{2} & -\frac{i}{\sqrt{2}} & \frac{1}{2} \\ -\frac{i}{2} & 0 & -\frac{i}{\sqrt{2}} \\ \frac{1}{2} & -\frac{i}{\sqrt{2}} & -\frac{1}{2} \end{pmatrix}$$

$$x = -1$$

$$D(T) = \begin{pmatrix} -i & 0 & 0 \\ 0 & -1 & 0 \\ 0 & 0 & i \end{pmatrix}$$

$$x = -1$$

$$D(U) = \begin{pmatrix} -\frac{1}{2} & -\frac{1}{\sqrt{2}} & -\frac{1}{2} \\ \frac{1}{\sqrt{2}} & 0 & -\frac{1}{\sqrt{2}} \\ -\frac{1}{2} & \frac{1}{\sqrt{2}} & -\frac{1}{2} \end{pmatrix}$$

$$x = -1$$

$$D(V) = \begin{pmatrix} -\frac{i}{2} & -\frac{i}{\sqrt{2}} & -\frac{i}{2} \\ -\frac{1}{\sqrt{2}} & 0 & \frac{1}{\sqrt{2}} \\ \frac{1}{2} & -\frac{i}{\sqrt{2}} & \frac{i}{2} \end{pmatrix}$$

$$x = 0$$

$$D(W) = \begin{pmatrix} -\frac{i}{2} & -\frac{1}{\sqrt{2}} & \frac{i}{2} \\ -\frac{i}{\sqrt{2}} & 0 & -\frac{i}{\sqrt{2}} \\ -\frac{i}{2} & \frac{1}{\sqrt{2}} & \frac{i}{2} \end{pmatrix}$$

$$\chi = 0$$

$$D(X) = \begin{pmatrix} -\frac{1}{2} & \frac{1}{\sqrt{2}} & -\frac{1}{2} \\ -\frac{1}{\sqrt{2}} & 0 & \frac{1}{\sqrt{2}} \\ -\frac{1}{2} & -\frac{1}{\sqrt{2}} & -\frac{1}{2} \end{pmatrix}$$

$$\chi = -1$$

$$D(E) = \begin{pmatrix} 1 & 0 & 0 \\ 0 & 1 & 0 \\ 0 & 0 & 1 \end{pmatrix}$$

The trace of the matrix corresponding to an element in an irreducible representation is the *character of the element in that representation*. We notice that all the elements of a given class have the same character in an irreducible representation. The characters are usually given classwise, as in Table II.

Two important theorems in representation theory are summarized in the formulas

$$\sum_R D_{ie}^{(\mu)}(R)\, D_{jm}^{(\nu)*}(R) = \frac{g}{n_\mu}\,\delta_{\mu\nu}\,\delta_{ij}\,\delta_{lm}$$

$$\sum_R \chi^{\mu}(R)\, \chi^{(\nu)*}(R) = g\,\delta_{\mu\nu} \qquad (9)$$

or equivalently,

TABLE II. CHARACTER TABLE OF S_4

	$K_1(=1^4)$	$K_2(21^2)$	$K_3(2^2)$	$K_4(31)$	$K_5(4)$
$D^{(1)} \equiv D^{(1^4)}$	1	-1	1	1	-1
$D^{(2)} \equiv D^{(21^2)}$	3	-1	-1	0	1
$D^{(3)} \equiv D^{(2^2)}$	2	0	2	-1	0
$D^{(4)} \equiv D^{(31)}$	3	1	-1	0	-1
$D^{(5)} \equiv D^{(4)}$	1	1	1	1	1

$$\sum_{i=1}^{k} X_i^{(\mu)} X_i^{(\nu)*} g_i = g\,\delta_{\mu\nu} \tag{10}$$

μ,ν are labels for the particular irreducible representation; $D_{ie}^{(\mu)}(R)$ is the ie-th matrix element of the matrix representing the group element O_R in the μ-th representation; n_μ is the dimension of the representation; $X_{(R)}^{(\mu)}$ is the character of the element O_R in the μ-th representation; $X_i^{(\mu)}$ is the character of the i-th class of elements (g_i in number) in the μ-th representation; and g is the order of the group. In the multiplication table for S_4, O_R stands for the elements E, A, B, C, D, and so forth, R being the running index. To demonstrate the relation (5), let us take $\mu = 2$, $\nu = 4$, $i = 1$, $l = 1$, $j = 3$, $m = 2$. From the matrices of the relevant irreducible representation we see, for instance,

$$D_{11}^{(2)}(A)\, D_{32}^{(4)*}(A) = \left(-\frac{i}{2}\right)\left(-\frac{i}{\sqrt{2}}\right) = -\frac{1}{2\sqrt{2}}$$

Summing appropriate products over all elements, we readily get

$$\left(-\frac{1}{2}\right)\left(-\frac{i}{\sqrt{2}}\right) + \left(\frac{i}{2}\right)\left(-\frac{1}{\sqrt{2}}\right) + (0)(0) + \left(-\frac{1}{2}\right)\left(-\frac{i}{\sqrt{2}}\right) + (1)(0) +$$

$$\left(-\frac{1}{2}\right)\left(-\frac{1}{\sqrt{2}}\right) + (0)(0) + \left(-\frac{1}{2}\right)\left(\frac{i}{\sqrt{2}}\right) + \left(\frac{i}{2}\right)\left(\frac{i}{\sqrt{2}}\right) + \left(-\frac{i}{2}\right)\left(-\frac{1}{\sqrt{2}}\right) +$$

$$(-1)(0) + (0)(0) + (0)(0) + \left(\frac{i}{2}\right)\left(-\frac{i}{\sqrt{2}}\right) + \left(\frac{i}{2}\right)\left(\frac{1}{\sqrt{2}}\right) + (-i)(0) +$$

$$\left(-\frac{1}{2}\right)\left(\frac{1}{\sqrt{2}}\right) + \left(\frac{1}{2}\right)\left(-\frac{i}{\sqrt{2}}\right) + \left(\frac{1}{2}\right)\left(\frac{i}{\sqrt{2}}\right) + (i)(0) + \left(\frac{1}{2}\right)\left(\frac{1}{\sqrt{2}}\right) + \left(-\frac{i}{2}\right)$$

$$\left(\frac{i}{\sqrt{2}}\right) + \left(-\frac{i}{2}\right)\left(\frac{1}{\sqrt{2}}\right) + \left(\frac{1}{2}\right)\left(-\frac{1}{\sqrt{2}}\right) = 0 \qquad (11)$$

If, on the other hand, we choose $\mu = 2 = \nu$, $i = j = \ell = m = 1$ the nonvanishing products add up to $24/3 = 8$ as follows:

$$\left(-\frac{i}{2}\right)\left(\frac{i}{2}\right) + \left(\frac{i}{2}\right)\left(-\frac{i}{2}\right) + (0)(0) + \left(-\frac{1}{2}\right)\left(-\frac{1}{2}\right) + (1)(1) + \left(-\frac{1}{2}\right)$$

$$\left(-\frac{1}{2}\right) + (0)(0) + \left(-\frac{1}{2}\right)\left(-\frac{1}{2}\right) + \left(\frac{i}{2}\right)\left(-\frac{i}{2}\right) + \left(-\frac{i}{2}\right)\left(\frac{i}{2}\right) + (-1)$$

$$(-1) + (0)(0) + (0)(0) + \left(\frac{i}{2}\right)\left(-\frac{i}{2}\right) + \left(\frac{i}{2}\right)\left(-\frac{i}{2}\right) + (-i)(i) +$$

$$\left(-\frac{1}{2}\right)\left(-\frac{1}{2}\right) + \left(\frac{1}{2}\right)\left(\frac{1}{2}\right) + \left(\frac{1}{2}\right)\left(\frac{1}{2}\right) + (i)(-i) + \left(\frac{1}{2}\right)\left(\frac{1}{2}\right) + \left(-\frac{i}{2}\right)\left(\frac{i}{2}\right) +$$

$$\left(-\frac{i}{2}\right)\left(\frac{i}{2}\right) + \left(\frac{1}{2}\right)\left(\frac{1}{2}\right) = \frac{24}{3} = 8 \qquad (12)$$

To demonstrate the orthogonality relations of characters, let $\mu = 2$, and $\nu = 4$. We then have from Table II, the character table.

$$\{1 \times 3 \times 3\} + \{-1 \times 1 \times 6\} + \{(-1) \times (-1) \times 3\} + \{8 \times o \times o\}$$

$$+ \{6 \times 1 \times -1\} = 9 - 6 + 3 + o - 6 = o$$

$\mu = 2 \quad \nu = 2$

$$\{1 \times 3 \times 3\} + \{(-1) \times (-1) \times 6\} + \{(-1) \times (-1) \times 3\}$$

$$+ \{o \times o \times 8\} + \{1 \times 1 \times 6\} = 9 + 6 + 3 + 6 = 24 = g \tag{13}$$

g = order of the group.

This theorem on characters helps to determine unknown characters. Suppose that b, c, d, f, k are unknown in the following character table of the tetrahedral group of rotations. Clearly, the first row has to be filled by four 1 s and the first column by numbers equal to the dimension of the representation, because the identity element represented by a n x n unit matrix in an n-dimensional irreducible representation.

	K_1 $g_1 = 1$	K_2 $g_2 = 3$	K_3 $g_3 = 4$	K_4 $g_4 = 4$
$\chi^{(1)}$	1	1	1	1
$\chi^{(2)}$	1	1	b	c
$\chi^{(3)}$	1	1	d	f
$\chi^{(4)}$	3	-1	o	k

The following algebraic calculations are straightforward:

$$\sum_i g_i \chi_i^{(1)}(R) \chi_i^{(2)*}(R) = (1 \times 1 \times 1) + (3 \times 1 \times 1)$$

$$+ (4 \times 1 \times b^*) + (4 \times 1 \times c^*) = 0 \tag{14}$$

or

$$4b^* + 4c^* = -4 \qquad b^* + c^* = -1$$

$$\sum_i g_i \chi_i^{(1)} \chi_i^{(4)*} = 3 - 3 + 0 + 4k^* = 0$$

or

$$K^* = 0 \therefore k = o$$

$$\sum_i g_i \chi_i^{(2)} \chi_i^{(2)*} = g = 12 = (1 \times 1) + (3 \times 1) + (4 \times b \times b^*)$$

$$+ (4 \times c \times c^*) \quad \text{or} \quad bb^* + cc^* = 2 \tag{15}$$

$$\sum_i g_i \chi_i^{(2)} \chi_i^{(3)*} = 1 + 3 + 4\,bd^* + 4\,cf^* = 0$$

$$\text{or } bd^* + cf^* = -1 \tag{16}$$

$$\sum_i g_i \chi_i^{(3)} \chi_i^{(3)*} = g = 12 = 1 + 3 + 4\,dd^* + 4\,ff^*$$

$$\text{or } dd^* + ff^* = 2 \tag{17}$$

$$\sum_i g_i \chi_i^{(1)} \chi_i^{(3)*} = 1 + 3 + 4\,d^* + 4\,f^* = 0$$

$$\text{or } d^* + f^* = -1 \tag{18}$$

Taking complex conjugates of all Equations (9) to (13) and collecting them we have

$$b^* + c^* = -1 \qquad b + c = -1$$

$$bb^* + cc^* = 2$$

$$bd^* + cf^* = -1 \qquad b^*d + c^*f = -1$$

$$dd^* + ff^* = 2$$

$$d^* + f^* = -1 \qquad d + f = -1 \tag{19}$$

Straightforward algebraic manipulation gives the solutions

$$f = \varepsilon = -\frac{1}{2} - i\,\frac{\sqrt{3}}{2} = e^{-\frac{2\pi i}{3}} = b$$

$$c = d = f^* = \varepsilon^* = \varepsilon^2 \tag{20}$$

It is well known that complex characters generally occur in pairs, and this completes the character table.

OTHER METHODS OF DETERMINING CHARACTERS

Another relation useful in calculating characters is the identity

$$K_i K_j = \sum_\ell c_{ij\ell} K_l \tag{21}$$

where K_i are classes and $c_{ij\ell}$ positive integers or zero. From this follows a similar relation for the characters of the classes in an irreducible representation:

$$g_i g_i \chi_i^{(\mu)} \chi_j^{(\mu)} = n_\mu \sum_\ell c_{ij\ell} g_\ell \chi_\ell^{(\mu)} \tag{22}$$

Here g_i is the number of elements in the class K_i, n_μ the dimension of the μ-th irreducible representation. To illustrate the first identity let us take classes K_3 and K_2 of S_4.

$$K_3 K_2 = (K,L,M) \times (C,D,F,G,H,Q)$$

$$= ((G,S,U,C,R,X),\ (T,H,X,P,D,U),\ (P,R,Q,T,S,F))$$

$$= K_2 + 2 K_5 = \sum_\ell c_{ij\ell} K_1,$$

$$c_{321} = 0 \quad c_{322} = 1 \quad c_{323} = 0 \quad c_{324} = 0$$

$$c_{325} = 2 \tag{23}$$

If we take the fourth representation the relation between the corresponding characters is easily demonstrated:

$$\mu = 4 \qquad g_3 = 3 \qquad g_2 = 6 \qquad n_\mu = 3$$

$$g_3 g_2 \chi_3^{(4)} \chi_2^{(4)} = 3 \times 6 \times (-1) \times 1 = -18$$

$$n_4 \sum_\ell c_{ij\ell} g_\ell \chi_\ell^{(4)} = 3 \{(o \times 1 \times 3) + (1 \times 6 \times 1) + (o \times 3 \times -1) + (o \times 8 \times o) + (2 \times 6 \times -1)\}$$

$$= 3(-6) = -18 \qquad (24)$$

It is interesting to note that a similar relation connects the number of elements in different classes.

That there ought to be a relation between the representations of a group and those of its subgroups is easily seen. In the group S_4 the matrices corresponding to E, K, L, M, A, B, I, J, N, O, V, W form a representation of the subgroup of order h = 12, although this need not be irreducible. This reducible representation of the subgroup can be decomposed into its irreducible components. Conversely, the characters of the subgroup help to build a compound character of the main group. To illustrate the prescription for obtaining these compound characters we write the classes of both S_4 and T:

S_4		T
K_1 (E)	$g_1 = 1$	$K_{1_1} = E$ $h_{1_1} = 1$; $h_1 = \sum_\tau h_{1_\tau}$ all other $h_{1_\tau} = 0$; $h_1 = 1$
K_2 (C,D,F,G,H,Q)	$g_2 = 6$	$K_{2_1} = 0$ all $h_{2_\tau} = 0$
K_3 (K,L,M)	$g_3 = 3$	$K_{3_1} = K,L,M$ $h_{3_1} = 3$ all other $h_1 = 3$ $h_{3_\tau} = 0$
K_4 (A,B,I,J,N,O,V,W)	$g_4 = 8$	$K_{4_1} = B,J,O,W$ $h_{4_1} = 4$
		$K_{4_2} = A,I,N,V$ $h_{4_2} = 4$
		$h_4 = h_{4_1} + h_{4_2} + h_{4_3} + h_{4_4}$ $= 4 + 4 + 0 + 0$ $= 8$
K_5 (P,R,S,T,U,X)	$g_5 = 6$	$K_{5_1} = 0$ all $h_{5_\tau} = 0$ $h_5 = 0$

(25)

The character table of T, the subgroup of order 12, is given in Table III.

TABLE III. CHARACTER TABLE OF T

	K_{1_1} (E)	K_{3_1} (K,L,M)	K_{4_1} (B,J,O,W)	K_{4_2} (A,I,N,V)
$D^{(1)}$	1	1	1	1
$D^{(2)}$	1	1	ε	ε^2
$D^{(3)}$	1	1	ε^2	ε
$D^{(4)}$	3	-1	0	0

Substituting the quantities in Table III in the formula

$$\psi_\ell^{(\mu)} = \sum_{\ell_\tau} \frac{h_{\ell_\tau}}{h} \frac{g}{g_\ell} \phi_{\ell_\tau}^{(\mu)} \tag{26}$$

appropriately, we obtain the compound characters $\psi_\ell^{(\mu)}$. For instance,

$$\frac{h_{4_1}}{h} = \frac{4}{12} = \frac{1}{3} \qquad \frac{g}{g_4} - 3 \qquad \phi_{4_1}^{(2)} = \varepsilon$$

$$\frac{h_{4_2}}{h} = \frac{4}{12} = \frac{1}{3} \qquad \frac{g}{g_4} = 3 \qquad \phi_{4_2}^{(2)} = \varepsilon^2 \tag{27}$$

$$\sum_{\ell_\tau} \frac{h_{\ell_\tau}}{h} \frac{g}{g_4} \phi_{4_\tau}^{(2)} = (\frac{1}{3} \times 3 \times \varepsilon) + (\frac{1}{3} \times 3 \times \varepsilon^2) = \varepsilon + \varepsilon^2 = -1$$

The compound character of class $\ell = 4$ (K_4) in the $\mu = 2$ representation is thus - 1. The other compound characters are obtained similarly. This character table is shown in Table IV.

TABLE IV. COMPOUND CHARACTER TABLE OF T

	K_1 (E) $g_1 = 1$	K_2 (K,L,M) $g_2 = 3$	K_3 (A,B,I,J,O, N,V,W) $g_3 = 8$	K_4 (C,D,F, G,H,Q) $g_4 = 6$	K_5 (P,R,S, T,U,X) $g_5 = 6$
$\psi^{(1)}$	2	0	2	2	0
$\psi^{(2)}$	2	0	2	-1	0
$\psi^{(3)}$	2	0	2	-1	0
$\psi^{(4)}$	6	0	-2	0	0

The reduction of these compound characters into simple characters is elaborated in Hamermesh's (1962) book. We indicate here a somewhat alternative procedure that works just as well in this case. From each of the compound characters $\psi^{(\mu)}$ if we subtract 1, which should be the character of every element in the symmetric representation $D^{(5)}$ of S_4, we get

	K_1	K_2	K_3	K_4	K_5
$\chi^{(1)}$	1	-1	1	1	-1 .

Since the character of the identity element $E(=K_1)$ is 1, it is obvious this is a set of simple characters in a one-dimensional irreducible representation. This representation is labeled $D^{(1)}$. $\psi^{(2)}$ satisfies the relation

$$\sum_i g_i \psi_i^{(2)^2} = (1 \times 2^2) + (6 \times 0^2) + (3 \times 2^2) + [8 \times (-1)^2] + (6 \times 0^2) = 24 \qquad (28)$$

and this shows that $\psi^{(2)}$ is also a set of simple characters of S_4, which belongs to the representation labeled $D^{(3)}$. From the theorem on the dimensions of the irreducible representations we notice the remaining two representations ought to be three dimensional:

$$1^2 + 1^2 + 2^2 + 3^2 + 3^2 = 24 .$$

A look at the compound character of the identity element in $\psi^{(4)}$ reveals that this ought to be a simple sum of the last two representations, because in a three-dimensional representation the identity element has character 3. Filling in algebraic unknowns for the remaining characters, we have the tentative character table

	$g_1 = 1$ K_1	$g_2 = 6$ K_2	$g_3 = 3$ K_3	$g_4 = 8$ K_4	$g_5 = 6$ K_5
$\chi^{(5)}$	1	1	1	1	1
$\chi^{(1)}$	1	-1	1	1	-1
$\chi^{(3)}$	2	0	2	-1	0
$\chi^{(2)}$	3	a	b	c	d
$\chi^{(4)}$	3	e	f	g	h

$\chi^{(2)} + \chi^{(4)} = \psi^{(4)}$ gives us readily

$$a + e = 0, \quad B + f = -2, \quad c = g = 0, \quad d + h = 0$$

From the fact that the sum of the (absolute) squares of the characters of all the elements in any irreducible representation equals the order of the group we see

$$9 + 6a^2 + 3b^2 + 8c^2 + 6d^2 = 24$$

$$9 + 6e^2 + 3f^2 + 8g^2 + 6h^2 = 24$$

From the orthogonality theorem on characters in different irreducible representations we have

$6 + (3 \times 2 \times b) + (8 \times -1 \times c) = 0$ (orthogonality of $\chi^{(2)}$ and $\chi^{(3)}$)

$6 + (3 \times 2 \times f) + (8 \times -1 \times b) = 0$ (orthogonality of $\chi^{(3)}$ and $\chi^{(4)}$)

$(1 \times 3) + (6 \times -1 \times e) + (3 \times 1 \times f) + (8 \times 1 \times g) + (6 \times -1 \times h) = 0$ (orthogonality of $\chi^{(1)}$ and $\chi^{(4)}$)

$(1 \times 3) + (6 \times 1 \times a) + (3 \times 1 \times b) + (8 \times 1 \times c) + (6 \times 1 \times d) = 0$ (orthogonality of $\chi^{(2)}$ and $\chi^{(5)}$)

A simple algebraic manipulation gives the solutions

$a = -1$	$e = +1$
$b = -1$	$f = -1$
$c = 0$	$g = 0$
$d = +1$	$h = -1$

and this completes the character table.

The most straightforward, though tedious, way of obtaining simple characters for any symmetric group is the application of the Frobenius formula. For the group S_3 and for a given partition and associated cycle structure this is

$$S_{(\ell)}D \equiv [s_1^{\alpha}\, s_2^{\beta}\, s_3^{\gamma}][\prod_{i<j}(x_i - x_j)]$$

$$= \sum_{(\lambda)} \chi^{(\lambda)}_{(\ell)} \sum_{P} \delta_P\, P\left(x_1^{\lambda_1+3-1}\; x_2^{\lambda_2+3-2}\; x_3^{\lambda_3}\right)$$

where

$$s_1^{\alpha}s_2^{\beta}s_3^{\gamma} = \left(\sum_{i=1}^{3} x_i\right)^{\alpha}\left(\sum_{i=1}^{3} x_i^2\right)^{\beta}\left(\sum_{i=1}^{3} x_i^{3}\right)^{\gamma} \qquad (29)$$

$$\pi(x_i - x_j) = (x_1 - x_2)\ (x_2 - x_3)\ (x_1 - x_3)$$

$$\sum_P \delta_P\ P(ABC) = ABC - BAC + BCA - CBA + CAB - ACB$$

It is to be noted that the expansion is for every class, whereas the summation is essentially over all representations. This formula then yields a column in the character table. We now briefly go through the details of the calculation for the characters of the irreducible representations of S_3.

Class $K_1 \to (3^1)$ $\alpha=0$ $\beta=0$ $\gamma=1$ $\lambda_1=3$ $\lambda_2=0$ $\lambda_3=0$

Class $K_2 \to (21)$ $\alpha=1$ $\beta=1$ $\gamma=0$ $\lambda_1=2$ $\lambda_2=1$ $\lambda_3=0$

Class $K_3 \to (1^3)$ $\alpha=3$ $\beta=0$ $\gamma=0$ $\lambda_1=1$ $\lambda_2=1$ $\lambda_3=1$

$$\lambda_1 \geq \lambda_2 \geq \lambda_3$$

$$\alpha + 2\beta + 3\gamma = 3 \tag{30}$$

Let us take the class K_2 corresponding to the partition $3 = 2+1 \to (21)$ $\pi(x_i - x_j) = (x_1^2\,x_2 - x_2^2x_1) + (x_2^2x_3 - x_3^2x_2) + (x_3^2x_1 - x_1^2x_3)$

$$s_1^\alpha s_2^\beta s_3^\gamma = (\Sigma\ x_i)^1 (\Sigma\ x_i^2)^1 (\Sigma\ x_i^3)^0 = (x_1+x_2+x_3)\ (x_1^2+x_2^2+x_3^2)$$

$$= (x_1^3+x_2^3+x_3^3) + \{(x_1^2x_2+x_2^2x_1) + (x_2^2x_3+x_3^2x_2) + (x_3^2x_1+x_1^2x_3)\} \tag{31}$$

Therefore,

$$s_{(\ell)}D = s_1^\alpha s_2^\beta s_3^\gamma\ (x_1 - x_2)\ (x_2 - x_3)\ (x_1 - x_3)$$

$$= \{(x_1^5x_2 - x_2^5x_1) + (x_2^5x_3 - x_3^5x_2) + (x_3^5x_1 - x_1^5x_3)\}$$

$$- \{(x_1^3x_2^2x_3 - x_2^3x_1^2x_3) + (x_2^3x_3^2x_1 - x_3^3x_2^2x_1) + (x_3^3x_1^2x_2 - x_1^3x_3^2x_2)\} \tag{32}$$

In the summation on the right side of the Equation (20) we notice that for the representation corresponding to $\{S\} \rightarrow \lambda_1=3\ \lambda_2=0=\lambda_3$ we should look for the coefficient of the term

$$x_1^{\lambda_1+3-1}\quad x_2^{\lambda_2+3-2}\quad x_3^{\lambda_3} = x_1^5 x_2\ , \tag{33}$$

whereas in the expansion all the permutations of terms like $x_1^5\ x_2$ occur with minus signs for odd permutations. The coefficient of this set of terms being + 1 the character of the class (21) in this representation is then + 1. Similarly, for the representation $\{\lambda\} \rightarrow \lambda_1 = 2,\ \lambda_2 = 1,\ \lambda_3 = 0$ the coefficient of the term

$$x_1^{2+3-1}\quad x_2^{1+3-2}\quad x_3^{o} = x_1^4 x_2^2 \tag{34}$$

and all its appropriate permutations is 0. The corresponding character is 0. For the representation $\{\lambda\} \rightarrow \lambda_1 = 1,\ \lambda_2 = 1,\ \lambda_3 = 1$ the coefficient of

$$x_1^3\ x_2^2\ x_3$$

and its permutations is - 1. Thus we have part of the character table

Representation	Class (21)
$D^{(3)}$	1
$D^{(21)}$	0
$D^{(1^3)}$	-1

(35)

The other characters can likewise be calculated from the expansions corresponding to the other two classes, and we arrive at the character table of Table V.

The characters of several groups like D_{3h}, C_{3h}, O_h, which are direct products of two groups, can be calculated easily from the characters of the factors. If $\chi^{(\mu)}(R_1)$ is the character of element O_{R_1} of one of the two groups in the μ-th irreducible representation and $\chi^{(\nu)}(R_2)$ the character of element O_{R_2} of the other factor in the ν-th irreducible representation of the latter, then their product

$$\chi^{\mu}(R_1)\ \chi^{(\nu)}(R_2) = \chi^{(\mu x \nu)}(R_1 R_2) \tag{36}$$

TABLE V. CHARACTER TABLE OF S_3 (ISOMORPHIC WITH C_{3v} AND D_3)

	K_1 1^3 $g_1 = 1$	K_2 (21) $g_2 = 3$	K_3 (3) $g_3 = 2$
$D^{(3)}$	1	1	1
$D^{(1^3)}$	1	-1	1
$D^{(21)}$	2	0	-1

is the character of the element of the direct product group corresponding to $O_{R_1} O_{R_2}$ in the $\mu x \nu$-th irreducible representation of this direct product. Let us take the group $T_h = C_i \otimes T$. The character tables of C_i (Table VI) and T (Table VII) specify the different classes and the elements in each class.

TABLE VI. CHARACTER TABLE OF C_i

	K_1 (E) $g_1 = 1$	K_2 (I) $g_2 = 1$
$D^{(1)}$	1	1
$D^{(2)}$	1	-1

TABLE VII. CHARACTER TABLE OF GROUP T

	K'_1 $g'_1 = 1$	K'_2 $g'_2 = 3$	K'_3 $g'_3 = 4$	K'_4 $g'_4 = 4$
$D^{(1')}$	1	1	1	1
$D^{(2')}$	1	1	ε	ε^2
$D^{(3')}$	1	1	ε^2	ε
$D^{(4')}$	3	-1	0	0

$$\varepsilon \equiv e^{-\frac{2\pi i}{3}}$$

The classes of the direct product group are

$$\begin{array}{ll} K_1 K'_1 \equiv C_1 \ (g_1 = 1) & K_2 K'_1 \equiv C_5 \ (g_5 = 1) \\ K_1 K'_2 \equiv C_2 \ (g_2 = 3) & K_2 K'_2 \equiv C_6 \ (g_6 = 3) \\ K_1 K'_3 \equiv C_3 \ (g_3 = 4) & K_2 K'_3 \equiv C_7 \ (g_7 = 4) \\ K_1 K'_4 \equiv C_4 \ (g_4 = 4) & K_3 K'_3 \equiv C_8 \ (g_8 = 4) \end{array} \quad (37)$$

From the product formula for characters we see, for instance,

$$\chi^{(1)}(K_1)\,\chi^{(4')}(K'_1) = 1 \times 3 = 3$$

$$\chi^{(2)}(K_1)\,\chi^{(2')}(K'_4) = 1 \times \varepsilon^2 = \varepsilon^2$$

$$\chi^{(2)}(K_2)\,\chi^{(3')}(K'_3) = -1 \times \varepsilon^2 = -\varepsilon^2 \quad (38)$$

Thus 3, ε^2, and $-\varepsilon^2$ are, respectively, the characters of the classes C_1, C_4, and C_7 in the (1 x 4'), (2 x 2'), and (2 x 3') irreducible representations. If we make all such multiplications, the complete character table of T_h is as given in Table VIII.

Regular Representation

From the multiplication table for the C_{3v} group of order 6, which is isomorphic to the symmetric group S_3 and the dihedral group D_3, we know, for instance, that DA = C. If we consider the result as a linear combination of all the elements we get

$$DA = 0E + 0A + 0B + 1C + 0D + 0F$$

and similarly

$$\begin{aligned} DE &= 0E + 0A + 0B + 0C + 1D + 0F \\ DB &= 0E + 1A + 0B + 0C + 0D + 0F \\ DC &= 0E + 0A + 1B + 0C + 0D + 0F \\ DD &= 0E + 0A + 0B + 0C + 0D + 1F \\ DF &= 1E + 0A + 0B + 0C + 0D + 0F \end{aligned} \tag{39}$$

If we write this correspondence in matrix form and then take the transpose of this matrix, we obtain what is known as the regular representation of D. Thus

$$D(D) = \begin{pmatrix} 0 & 0 & 0 & 0 & 0 & 1 \\ 0 & 0 & 1 & 0 & 0 & 0 \\ 0 & 0 & 0 & 1 & 0 & 0 \\ 0 & 1 & 0 & 0 & 0 & 0 \\ 1 & 0 & 0 & 0 & 0 & 0 \\ 0 & 0 & 0 & 0 & 1 & 0 \end{pmatrix} \tag{40}$$

The matrices representing the other elements in the regular representation are likewise

TABLE VIII. CHARACTER TABLE OF T_h

	C_1 $g_1 = 1$	C_2 $g_2 = 3$	C_3 $g_3 = 4$	C_4 $g_4 = 4$	C_5 $g_5 = 1$	C_6 $g_6 = 3$	C_7 $g_7 = 4$	C_8 $g_8 = 4$
$D^{(1x1')} \equiv A_g$	1	1	1	1	1	1	1	1
$D^{(1x2')} \equiv E_g$	1	1	ε	ε^2	1	1	ε	ε^2
$D^{(1x3')} \equiv E_g$	1	1	ε^2	ε	1	1	ε^2	ε
$D^{(1x4')}$	3	-1	0	0	3	-1	0	0
$D^{(2x1')} \equiv A_u$	1	1	1	1	-1	-1	-1	-1
$D^{(2x2')} \equiv E_u$	1	1	ε	ε^2	-1	-1	$-\varepsilon$	$-\varepsilon^2$
$D^{(2x3')} \equiv E_u$	1	1	ε^2	ε	-1	-1	$-\varepsilon^2$	$-\varepsilon$
$D^{(2x4')}$	3	-1	0	0	-3	1	0	0

$$D(E) = \begin{pmatrix} 1&0&0&0&0&0 \\ 0&1&0&0&0&0 \\ 0&0&1&0&0&0 \\ 0&0&0&1&0&0 \\ 0&0&0&0&1&0 \\ 0&0&0&0&0&1 \end{pmatrix} \qquad D(A) = \begin{pmatrix} 0&1&0&0&0&0 \\ 1&0&0&0&0&0 \\ 0&0&0&0&1&0 \\ 0&0&0&0&0&1 \\ 0&0&1&0&0&0 \\ 0&0&0&1&0&0 \end{pmatrix}$$

$$D(B) = \begin{pmatrix} 0&0&1&0&0&0 \\ 0&0&0&0&0&1 \\ 1&0&0&0&0&0 \\ 0&0&0&0&1&0 \\ 0&0&0&1&0&0 \\ 0&1&0&0&0&0 \end{pmatrix} \qquad D(C) = \begin{pmatrix} 0&0&0&1&0&0 \\ 0&0&0&0&1&0 \\ 0&0&0&0&0&1 \\ 1&0&0&0&0&0 \\ 0&1&0&0&0&0 \\ 0&0&1&0&0&0 \end{pmatrix}$$

$$D(F) = \begin{pmatrix} 0&0&0&0&1&0 \\ 0&0&0&1&0&0 \\ 0&1&0&0&0&0 \\ 0&0&1&0&0&0 \\ 0&0&0&0&0&1 \\ 1&0&0&0&0&0 \end{pmatrix} \qquad (41)$$

In this representation we notice the character of all the elements, except E, happens to be 0 and the character of E equals the order of the group, 6 here. It is obvious these are compound characters and this is a reducible representation. Relating the compound characters to the characters of C_{3v} in the various irreducible representations we have

	K_1 (E)	K_2 (A,B,C)	K_3 (D,F)
χ	6	0	0
$\chi^{(1)}$	1	1	1
$\chi^{(2)}$	1	-1	1
$\chi^{(3)}$	2	0	-1

We readily see that $\chi = \chi^{(1)} + \chi^{(2)} + 2\chi^{(3)}$, or $D = D^{(1)} \oplus D^{(2)} \oplus 2D^{(3)}$. Thus in the regular representation, each irreducible representation occurs the number of times that are equal to the dimension of the irreducible representation. This fact has a bearing on the characteristic equation in the theory of algebras. There are as many irreducible factors in the characteristic determinant as there are commuting elements, and each factor, of degree n in the variables, is raised to the power n. The elements of the group, taken as basis elements, form an algebra called the Frobenius algebra. The identity element of the group E satisfies $E^2 = E$ and commutes with every element of the algebra:

$$Ex = xE = x$$

The algebra thus has a modulus as well as an idempotent, and it has no properly nilpotent element. This is, therefore, a semi-simple algebra which is equivalent to a direct sum of $n \times n$ matrix algebras. The number of such matrix algebras is equal to the number of distinct commuting elements of the algebra which number, in this special case of a Frobenius algebra, happens to be equal to the number of classes in the group because the sum of elements in any class commutes with every element of the algebra. The *sums* of these classes are themselves basis elements of another Abelian algebra. These were the results we demonstrated explicitly in Chapter One.

AN EXAMPLE OF CONSTRUCTION OF REPRESENTATIONS

To construct representations we have to start with a set of linearly independent basis functions. Let us take the Cartesian coordinates x, y, z as arguments of these basis functions. Before we make a choice of a set, let us see what the various symmetry operations, which are elements of the finite group C_{3v}, do to x, y, and z. This is easy to do geometrically if we choose the centroid of an equilateral triangle as the origin of a rectangular coordinate system. In Figure 1 the Y axis passes through vertex 1, and the Z axis is perpendicular to the plane of the paper, pointing upwards. In the case of the NH_3 molecule the three hydrogen atoms can be located at vertices 1, 2, and 3, with the nitrogen atom at some point on the upward Z axis. The coordinates of these three atoms are shown in parentheses.

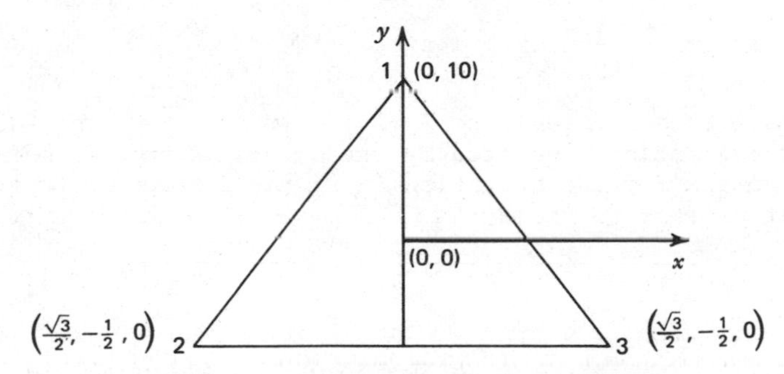

Figure 1. Coordinates of the vertices.

The positions of the H atoms after the respective operations of C_{3v} are shown in Figure 2.

1 2 △ 3	3 1 △ 2	2 3 △ 1	1 3 △ 2	3 2 △ 1	2 1 △ 3
E	C_3	C_3^2	σ_v	$\sigma_v, \sigma_{v''}$	$\sigma_v, \sigma_{v'}$

Figure 2. Results of operations $O_{R'}$.

The effect of the two cyclic rotations $O_R = C_3$ and $O_R = C_3^2$ on x, y, and z can be obtained from the matrix relating the components x', y', and z' of a vector rotated clockwise through an angle ψ about the Z axis and the original components x, y, and z (or components of the same vector rotated through 0° angle)

$$\begin{pmatrix} x' \\ y' \\ z' \end{pmatrix} = \begin{pmatrix} \cos\psi & -\sin\psi & 0 \\ \sin\psi & \cos\psi & 0 \\ 0 & 0 & 1 \end{pmatrix} \begin{pmatrix} x \\ y \\ z \end{pmatrix} \tag{42}$$

For the operation C_3, $\psi = 2\pi/3$ and for C_3^2, $\psi = 4\pi/3$. The effect of reflection in the YZ plane (i.e., the operation σ_v interchanges 3 and 2) or x goes to - x, with y and z remaining unchanged. The effect of $\sigma_{v''}$ is the same as rotating the triangle first through $2\pi/3$ and then reflecting in the YZ plane. Similarly, the operation $\sigma_{v'}$ is the same as the C_3^2 rotation followed by a reflection in the YZ plane. Let us remember that the coordinate system is held fixed while the positions of the atoms are interchanged according to the symmetry operations of the group. The results are collected in Table IX.

Let us choose the basis functions

$$\psi_1 = x + z, \qquad \psi_2 = y + z, \qquad \psi_3 = z \ . \tag{43}$$

Since we have chosen three functions, we will get a three-dimensional representation generated by them. To calculate the representation matrix for the operation σ_v we proceed as follows. From Table IX it is easy to verify

$$\sigma_v \, \psi_1 \equiv O_R \, \psi_1 = - x + z = - \psi_1 + 0 \, \psi_2 + 2\psi_3 \; [R \rightarrow \sigma_v]$$

$$= D_{11}(\sigma_v) \, \psi_1 + D_{21}(\sigma_v) \, \psi_2 + D_{31}(\sigma_v) \, \psi_3$$

$$\sigma_v \, \psi_2 = 0\psi_1 + 1\psi_2 + 0\psi_3$$

$$\sigma_v \, \psi_3 = 0\psi_1 + 0\psi_2 + 1\psi_3 \tag{44}$$

According to the formula

$$O_R\psi_\nu = \sum_{\mu=1}^{n} \psi_\mu \, D_{\mu\nu}(R) \tag{45}$$

d in this basis by the matrix σ_v is represented, in this basis, by the matrix

$$D(\sigma_v) = \begin{pmatrix} -1 & 0 & 0 \\ 0 & 1 & 0 \\ 2 & 0 & 1 \end{pmatrix}$$

Similarly, the other elements of the group are represented by the

TABLE IX. $O_R x$, $O_R y$, $O_R z$

	$E\ X_i$	F(132) $C_3\ X_i$	D(123) $C_3\ X_i$	A(23) $\sigma_v\ X_i$	C(12) $\sigma_v'\ X_i$	B(13) $\sigma_v''\ X_i$
$X_1 = X$	x	$-\frac{1}{2}x - \frac{\sqrt{3}}{2}y$	$-\frac{1}{2}x + \frac{\sqrt{3}}{2}y$	$-x$	$\frac{1}{2}x - \frac{\sqrt{3}}{2}y$	$\frac{1}{2}x + \frac{\sqrt{3}}{2}y$
$X_2 = Y$	y	$\frac{\sqrt{3}}{2}x - \frac{1}{2}y$	$-\frac{\sqrt{3}}{2}x - \frac{1}{2}y$	y	$-\frac{\sqrt{3}}{2}x - \frac{1}{2}y$	$\frac{\sqrt{3}}{2}x - \frac{1}{2}y$
$X_3 = Z$	z	z	z	z	z	z

following matrices:

$$D(E) = \begin{pmatrix} 1 & 0 & 0 \\ 0 & 1 & 0 \\ 0 & 0 & 1 \end{pmatrix} \quad D(C_3) = \begin{pmatrix} -\frac{1}{2} & \frac{\sqrt{3}}{2} & 0 \\ -\frac{\sqrt{3}}{2} & -\frac{1}{2} & 0 \\ \frac{3+\sqrt{3}}{2} & \frac{3-\sqrt{3}}{2} & 1 \end{pmatrix}$$

$$D(C_3^2) = \begin{pmatrix} -\frac{1}{2} & -\frac{\sqrt{3}}{2} & 0 \\ \frac{\sqrt{3}}{2} & -\frac{1}{2} & 0 \\ \frac{3-\sqrt{3}}{2} & \frac{3+\sqrt{3}}{2} & 1 \end{pmatrix} \quad D(\sigma_{v'}) = \begin{pmatrix} \frac{1}{2} & -\frac{\sqrt{3}}{2} & 0 \\ -\frac{\sqrt{3}}{2} & -\frac{1}{2} & 0 \\ \frac{1+\sqrt{3}}{2} & \frac{3+\sqrt{3}}{2} & 1 \end{pmatrix}$$

$$D(\sigma_{v''}) = \begin{pmatrix} \frac{1}{2} & \frac{\sqrt{3}}{2} & 0 \\ \frac{\sqrt{3}}{2} & -\frac{1}{2} & 0 \\ \frac{1-\sqrt{3}}{2} & \frac{3-\sqrt{3}}{2} & 1 \end{pmatrix} \tag{46}$$

The (compound) character table of the above representation (Γ say) now becomes

	E	C_3	C_3^2	σ_v	$\sigma_{v'}$	$\sigma_{v''}$	
Γ	3	0	0	1	1	1	(47)

Remembering (a) that C_{3v} does not have a three-dimensional irreducible representation according to the theorem relating the dimensionality of the representation to the order of the group, (b) the character table above is not the same as the well-known character table for C_{3v}, and (b) the matrices are not all in block diagonal form, we see that the basis $\phi_1\phi_2\phi_3$ generates a *reducible* representation. In other words, this is not the right basis set.

There then ought to exist a matrix S that, by a similarity transformation, simultaneously reduces all the above matrices of the reducible representation Γ into block diagonal form wherein each matrix can be expressed as a direct sum of (lower dimensional) simpler matrices. To find this matrix S we notice that the known two-dimensional irreducible representation of σ_v consists only of diagonal elements. The matrix that diagonalizes $D(\sigma_v)$ of Γ can be calculated following the methods of Chapter One. A straightforward calculation shows that

$$\underset{\sim}{S}\, D(\sigma_v)\underset{\sim}{S}^{-1} \equiv \begin{pmatrix} 0 & 1 & 0 \\ -1 & 0 & 0 \\ 1 & 1 & 1 \end{pmatrix} \begin{pmatrix} -\,0 & 0 & 0 \\ 0 & 1 & 0 \\ 2 & 0 & 1 \end{pmatrix} \begin{pmatrix} 0 & -1 & 0 \\ 1 & 0 & 0 \\ -1 & 1 & 1 \end{pmatrix} = \begin{pmatrix} 1 & 0 & 0 \\ 0 & -1 & 0 \\ 0 & 0 & 1 \end{pmatrix} \tag{48}$$

Now we try if S can put the other matrices of Γ in this form. A straightforward matrix multiplication gives the following results:

$$S\, D(C_3)S^{-1} = \begin{pmatrix} -\frac{1}{2} & \frac{\sqrt{3}}{2} & 0 \\ -\frac{\sqrt{3}}{2} & -\frac{1}{2} & 0 \\ 0 & 0 & 1 \end{pmatrix};$$

$$S\, D(C_3^2)S^{-1} = \begin{pmatrix} -\frac{1}{2} & -\frac{\sqrt{3}}{2} & 0 \\ \frac{\sqrt{3}}{2} & -\frac{1}{2} & 0 \\ 0 & 0 & 1 \end{pmatrix};$$

$$S\,D(\sigma_{v'})S^{-1} = \begin{pmatrix} -\frac{1}{2} & \frac{\sqrt{3}}{2} & 0 \\ \frac{\sqrt{3}}{2} & \frac{1}{2} & 0 \\ 0 & 0 & 1 \end{pmatrix};$$

$$S\,D(\sigma_{v''})S^{-1} = \begin{pmatrix} -\frac{1}{2} & -\frac{\sqrt{3}}{2} & 0 \\ -\frac{\sqrt{3}}{2} & \frac{1}{2} & 0 \\ 0 & 0 & 1 \end{pmatrix}. \tag{49}$$

The transformation of the identity element E is trivial. Thus we see that the S matrix does reduce all matrices, and the latter can be written as direct sums of two irreducible representation matrices, one two-dimensional and the other one-dimensional, as follows:

$$\begin{array}{cccc} & E & C_3 & C_3^2 \\ D^{(E)}: & \begin{pmatrix} 1 & 0 \\ 0 & 1 \end{pmatrix} & \begin{pmatrix} -\frac{1}{2} & \frac{\sqrt{3}}{2} \\ -\frac{\sqrt{3}}{2} & -\frac{1}{2} \end{pmatrix} & \begin{pmatrix} -\frac{1}{2} & -\frac{\sqrt{3}}{2} \\ \frac{\sqrt{3}}{2} & -\frac{1}{2} \end{pmatrix} \\ & X(E) = 2 & X(C_3) = -1 & X(C_3^2) = -1 \end{array}$$

$$\begin{array}{cccc} & \sigma_v & \sigma_{v'} & \sigma_{v''} \\ D^{(E)}: & \begin{pmatrix} 1 & 0 \\ 0 & -1 \end{pmatrix} & \begin{pmatrix} -\frac{1}{2} & \frac{\sqrt{3}}{2} \\ \frac{\sqrt{3}}{2} & \frac{1}{2} \end{pmatrix} & \begin{pmatrix} -\frac{1}{2} & -\frac{\sqrt{3}}{2} \\ -\frac{3}{2} & \frac{1}{2} \end{pmatrix} \\ & X(\sigma_v) = 0 & X(\sigma_{v'}) = 0 & X(\sigma_{v''}) = 0 \end{array}$$

$$D^{(A_1)} : \quad 1 \quad 1 \quad 1 \quad 1 \quad 1 \quad 1 \tag{50}$$

That the above two-dimensional representation is irreducible is easily seen by comparing the above characters with the known characters of C_{3v}. This comparison shows that Γ is a sum of the totally symmetric A_1 representation and the degenerate (E) representation of C_{3v}

$$\Gamma = D^{(E)} \oplus D^{(A_1)} \tag{51}$$

It is customary to say that in the reducible representation Γ, the irreducible representation A_1 occurs once and the irreducible representation E occurs once. This can also be inferred by inspection of the characters of the reducible and irreducible representations

$$\chi(\Gamma) = \chi(A_1) + \chi(E) \tag{52}$$

for all the elements of C_{3v}.

It is important to note that the basis set $\psi_1 = x$, $\psi_2 = y$ generates the following two-dimensional representation with the respective characters.

$$D(E) = \begin{pmatrix} 1 & 0 \\ 0 & 1 \end{pmatrix} \quad D(C_3^2) = \begin{pmatrix} -\frac{1}{2} & -\frac{\sqrt{3}}{2} \\ \frac{\sqrt{3}}{2} & -\frac{1}{2} \end{pmatrix} \quad D(C_3) = \begin{pmatrix} -\frac{1}{2} & \frac{\sqrt{3}}{2} \\ -\frac{\sqrt{3}}{2} & -\frac{1}{2} \end{pmatrix}$$

$$\chi(E) = 2 \qquad \chi = -1 \qquad \chi = -1$$

$$D(\sigma_v) = \begin{pmatrix} -1 & 0 \\ 0 & 1 \end{pmatrix} \quad D(\sigma_{v'}) = \begin{pmatrix} \frac{1}{2} & -\frac{\sqrt{3}}{2} \\ -\frac{\sqrt{3}}{2} & -\frac{1}{2} \end{pmatrix} \quad D(\sigma_{v''}) = \begin{pmatrix} \frac{1}{2} & \frac{\sqrt{3}}{2} \\ \frac{\sqrt{3}}{2} & -\frac{1}{2} \end{pmatrix}$$

$$\chi = 0 \qquad \chi = 0 \qquad \chi = 0 \tag{53}$$

$\psi = z$ likewise generates the following one-dimensional representation:

E	C_3	C_3^2	σ_v	$\sigma_{v'}$	$\sigma_{v''}$
1	1	1	1	1	1

One sure way of knowing whether the given representation is reducible or irreducible is by looking at the characters. If the characters are the same as those of the irreducible representation, the given representation differs from the known one only through a similarity transformation; in other words, the representation is equivalent to an irreducible representation. This is because of the theorem in matrix theory that the trace of a given matrix is invariant to a similarity transformation. By comparing the characters it is readily seen that (x,y) are basis functions, in that order, for a two-dimensional irreducible representation of C_{3v} and that z generates the totally symmetric A_1 irreducible representation. Alternatively, the pair (x,y) transforms as the E rperesentation of C_{3v} and z as the A_1 representation. These transformation properties are usually shown in character tables. x is said to belong to the first row of the irreducible representation E of C_{3v} and y to the second row of the same representation. The above set of matrices can be brought into coincidence with the set obtained by reduction of Γ, with the help of the unitary matrix U for the similarity transformation.

$$U = \begin{pmatrix} 0 & 1 \\ -1 & 0 \end{pmatrix} \tag{54}$$

Applying the formula

$$P_i^{(\mu)} = \frac{n_\mu}{g} \sum_R D_{ii}^{(\mu)*}(R) O_R \tag{55}$$

we notice that the following two operators are projection operators with respect to the above irreducible representation:

$$P_1 = \frac{2}{6} \sum_R D_{11}(R)\, O_R$$

$$= 1/3\ (E - \frac{1}{2} C_3 - \frac{1}{2} C_3^2 - \sigma_v + \frac{1}{2} \sigma_{v'} + \frac{1}{2} \sigma_{v''})$$

$$P_2 = 1/3 \sum_R D_{22}(R)\, O_R$$

$$= 1/2\ (E - \frac{1}{2} C_3 - \frac{1}{2} C_3^2 + \sigma_v - \frac{1}{2} \sigma_{v'} - \frac{1}{2} \sigma_{v''}) \qquad (56)$$

If we choose a single function $\phi(x,y,z) = x + y + z$, Table IX helps us to evaluate $P_1\phi$ and $P_2\phi$ with the results

$$P_1\phi = x$$

$$P_2\phi = y \qquad (57)$$

The projection operators thus serve to generate the basis functions of the irreducible representation starting from a single arbitrary function

$$\psi_i^{(\nu)} = P_i^{(\nu)}\ \phi \qquad (58)$$

CHAPTER FOUR

CONTINUOUS LIE GROUPS

STRUCTURE RELATIONS

Let us define a (Hermitian) linear transformation in two variables $x(\equiv x_1)$ and $y(\equiv x_2)$:

$$x' = a_1 x + (b_1 + ib_2)y$$

$$y' = (b_1 - ib_2)x + \frac{1 + b_1^2 + b_2^2}{a_1} y \qquad (1)$$

It is easy to see that the transformations, infinite in number, satisfy the group postulates if the parameters a_1, b_1, b_2 can take on all possible values in the field of real numbers and if by "group multiplication" we mean successive transformations. The "identity", for instance, is given by $a_1 = 1$, $b_1 = 0 = b_2$ (equilibrium values) and the inverse transformation by

$$a_1^{\text{inverse}} = \frac{1 + b_1^2 + b_2^2}{a_1}, \quad b_1^{\text{inverse}} = -b_1,$$

$$b_2^{\text{inverse}} = -b_2 .$$

When the range of values of the parameters is infinite, as in this case, it is known as a "noncompact group". In his fundamental work Sophus Lie (1893) pointed out that the finite transformations can be generated by means of "infinitesimal generators" and established the following relations:

$$\text{I.} \quad X_i' = \left(e^{\varepsilon_1 \underset{\sim}{X}_1 + \varepsilon_2 \underset{\sim}{X}_2 + \varepsilon_3 \underset{\sim}{X}_3} \right) X_i$$

$$\text{II.} \quad \underset{\sim}{X}_\alpha = \sum_{i=1}^{2} U_{i\alpha} \frac{\partial}{\partial X_i}, \quad U_{i\alpha} = \left. \frac{\partial X_i}{\partial \alpha} \right|_{\text{equilibrium}}$$

$$\text{III.} \quad [\underset{\sim}{X}_\alpha, \underset{\sim}{X}_\beta] = \sum_\gamma C^\gamma_{\alpha\beta} \underset{\sim}{X}_\gamma \quad (C^\gamma_{\alpha\beta} \text{ called structure constants}) \tag{2}$$

There are as many infinitesimal generators as there are independent parameters, and in our example these are labeled X_1, X_2, X_3 corresponding to α being a_1, b_1 and b_2. Thus we have

$$U_{11} = \frac{\partial X_1}{\partial a_1} = x, \quad \frac{\partial X_2}{\partial a_1} = \frac{\partial y}{\partial a_1} = -\left(\frac{1 + b_1^2 + b_2^2}{a_1^2} \right) y$$

$$= -y \text{ at the equilibrium values } a_1=1,\ b_1=0=b_2$$

$$\underset{\sim}{X}_1 = x \frac{\partial}{\partial x} - y \frac{\partial}{\partial y}$$

Similarly,

$$\underset{\sim}{X}_2 = y \frac{\partial}{\partial x} + x \frac{\partial}{\partial y}, \quad \underset{\sim}{X}_3 = i\left(y \frac{\partial}{\partial x} - x \frac{\partial}{\partial y}\right) \tag{3}$$

If $f(x,y)$ is an arbitrary function of x and y, a simple calculations shows

$$[\underset{\sim}{X}_1, \underset{\sim}{X}_2]f = \left(x \frac{\partial}{\partial x} - y \frac{\partial}{\partial y} \right) \left(y \frac{\partial f}{\partial x} + x \frac{\partial f}{\partial y} \right)$$

$$- \left(y \frac{\partial}{\partial x} + x \frac{\partial}{\partial y} \right) \left(x \frac{\partial f}{\partial x} - y \frac{\partial f}{\partial y} \right)$$

$$= 2\left(x \frac{\partial}{\partial y} - y \frac{\partial}{\partial x}\right) f = 2i \underset{\sim}{X}_3 f$$

$$\text{OR } [X_1, X_2] = 2i \underset{\sim}{X}_3 \tag{4}$$

The other commutators happen to be

$$[X_2, X_3] = 2i \underset{\sim}{X}_1, \quad [\underset{\sim}{X}_3, \underset{\sim}{X}_1] = 2i \underset{\sim}{X}_2 .$$

The structure constants are then

$$C^3_{12} = 2i = - C^3_{21}, \; C^1_{23} = 2i = - C^1_{32}, \; C^2_{13} = - 2i = - C^2_{31} \tag{5}$$

All other $C^{\gamma}_{\alpha\beta}$ (21 of them) vanish. To establish that the finite transformation can be generated with the X_i operators one needs to use

$$e^{\Sigma\varepsilon_i \underset{\sim}{X}_i} = 1 + \Sigma\varepsilon_i X_i + \frac{1}{2!} \Sigma\varepsilon_i X_i \; \Sigma\varepsilon_j X_j + \cdots \tag{6}$$

A straightforward calculations shows

$$x' = \left(e^{\varepsilon_1 \underset{\sim}{X}_1 + \varepsilon_2 \underset{\sim}{X}_2 + \varepsilon_3 \underset{\sim}{X}_3} \right) x = \left(\frac{\varepsilon_1}{\varepsilon} \sinh \varepsilon + \cosh \varepsilon \right) x$$

$$+ \left(\frac{\varepsilon_2 + i\varepsilon_3}{\varepsilon} \sinh \varepsilon \right) y$$

$$y' = \left(e^{\Sigma_i \varepsilon_i \underset{\sim}{X}_i} \right) y = \left[\frac{\varepsilon_2 - i\varepsilon_3}{\varepsilon} \sinh \varepsilon \right] x$$

$$+ \left[\frac{\varepsilon^2 + (\varepsilon_2^2 + \varepsilon_3^2) \sinh^2 \varepsilon}{\varepsilon \, \varepsilon_1 \sinh \varepsilon + \varepsilon^2 \cosh \varepsilon} \right] y$$

$$\text{where}\quad \varepsilon^2 = \varepsilon_1^2 + \varepsilon_2^2 + \varepsilon_3^2 \tag{7}$$

In other words, a_1, b_1, b_2 are functions of the parameters ε_1, ε_2, ε_3, all of which are assumed to be real. Campbell's book (1966) gives the appropriate relations between the ε s and the three Eulerian angles in the case of the three-dimensional rotation group.

SEMI-SIMPLE GROUPS, CASIMIR OPERATOR

A group is said to be semisimple if the determinant $||g_{\mu\nu}||$ does not vanish, $g_{\mu\nu}$ being defined by

$$g_{\mu\nu} = \sum_{\alpha\beta} C^{\beta}_{\mu\alpha} C^{\alpha}_{\nu\beta} \tag{8}$$

In our example we notice that $g_{\mu\nu}$ is in matrix form

$$g_{\mu\nu} = \begin{pmatrix} 8 & 0 & 0 \\ 0 & 8 & 0 \\ 0 & 0 & 8 \end{pmatrix} \quad \text{and} \quad ||g_{\mu\nu}|| = 512 \neq 0 \tag{9}$$

It is thus a semisimple group. An important property of semisimple groups is that there exists an invariant operator, called the Casimir operator, defined by

$$\underset{\sim}{C} = \sum_{\mu\nu} g^{\mu\nu} \underset{\sim}{X}_{\mu} \underset{\sim}{X}_{\nu} \qquad (g^{\mu\nu} = \text{inverse of } g_{\mu\nu}) \tag{10}$$

which commutes with all the infinitesimal generators of the group $[C, \underset{\sim}{X}_{\alpha}] = 0$. For the Lie group under discussion

$$g^{\mu\nu} = \frac{1}{8} \begin{pmatrix} 1 & 0 & 0 \\ 0 & 1 & 0 \\ 0 & 0 & 1 \end{pmatrix}$$

$$\underset{\sim}{C} = \frac{1}{8}\left[\underset{\sim}{X}_1^2 + \underset{\sim}{X}_2^2 + \underset{\sim}{X}_3^2\right]$$

$$= \frac{1}{8}\left(x\frac{\partial}{\partial x} + y\frac{\partial}{\partial y}\right)\left(x\frac{\partial}{\partial x} + y\frac{\partial}{\partial y} + 3\right) \tag{11}$$

CHAPTER FIVE

APPLICATIONS IN ATOMIC PHYSICS

SYMMETRY OF THE NONRELATIVISTIC HYDROGEN ATOM

The four-dimensional rotation group O(4) is the group of real orthogonal transformations that leave

$$x_1^2 + x_2^2 + x_3^2 + x_4^2$$

invariant. This continuous Lie group has six parameters and six infinite simal generators:

$$L_1 = x_3 \frac{\partial}{\partial x_2} - x_2 \frac{\partial}{\partial x_3}$$

$$L_2 = x_1 \frac{\partial}{\partial x_3} - x_3 \frac{\partial}{\partial x_1}$$

$$L_3 = x_2 \frac{\partial}{\partial x_1} - x_1 \frac{\partial}{\partial x_2}$$

$$A_1 = x_1 \frac{\partial}{\partial x_4} - x_4 \frac{\partial}{\partial x_1}$$

$$A_2 = x_2 \frac{\partial}{\partial x_4} - x_4 \frac{\partial}{\partial x_2}$$

$$A_3 = x_3 \frac{\partial}{\partial x_4} - x_4 \frac{\partial}{\partial x_3} \tag{1}$$

These satisfy the Lie algebra

$$[L_i, L_j] = \sum_k C^k_{ij} L_k = \sum_k \varepsilon_{ijk} L_k$$

$$[A_i, A_j] = \sum_k \varepsilon_{ijk} L_k$$

$$[L_i, A_j] = \sum_k \varepsilon_{ijk} A_k \tag{2}$$

The structure constants are given by the Levi-Civita tensor symbol ε_{ijk} which assumes the value O, whenever any two indices are equal, is + 1 when ijk is an even permutation of 123 and -1 when odd. For instance, $C^2_{12} = \varepsilon_{122} = 0$, $C^3_{12} = \varepsilon_{123} = +1$, $C^3_{21} = \varepsilon_{213} = -1$. This group has two invariants that commute with every infinitesimal generator

$$F = L_1^2 + L_2^2 + L_3^2 + A_1^2 + A_2^2 + A_3^2$$

$$G = L_1A_1 + L_2A_2 + L_3A_3 \tag{3}$$

$$\underset{\sim}{H} = \frac{1}{2\mu} \vec{p}^2 - e^2 \frac{1}{r} \qquad r = (x_1^2 + x_2^2 + x_3^2)^{\frac{1}{2}} \qquad \vec{p}^2 = (p_1^2 + p_2^2 + p_3^2) \tag{4}$$

With quantum conditions it can be shown that the orbital angular momentum operator $\vec{L}$ and the Runge-Lenz-Pauli operator $\vec{A}$ commute with the Hamiltonian and are, therefore, constants of the motion. For the discrete spectrum of $\underset{\sim}{H}$ these operators are explicitly (in units of $\hbar = c = 1$)

$$L_x = x_2p_3 - x_3p_2 = -i\, x_2 \frac{\partial}{\partial x_3} + i\, x_3 \frac{\partial}{\partial x_2}, \quad p_3 = -i \frac{\partial}{\partial x_3}$$

etc.

$$L_y = x_3 p_1 - x_1 p_3 = -i\, x_3 \frac{\partial}{\partial x_1} + i\, x_1 \frac{\partial}{\partial x_3}$$

$$L_z = x_1 p_2 - x_2 p_1 = -i\, x_1 \frac{\partial}{\partial x_2} + i\, x_2 \frac{\partial}{\partial x_1}$$

$$A_x = \sqrt{\frac{1}{8\mu E}} \left[(p_y L_z - p_z L_y) - (L_y p_z - L_z p_y)\right] - \sqrt{\frac{\mu}{2E}}\, \frac{x_1}{r}$$

$$A_y = \sqrt{\frac{1}{0\mu E}} \left[(p_z L_x - p_x L_z) - (L_z p_x - L_x p_z)\right] \quad \sqrt{\frac{\mu}{2E}}\, \frac{x_2}{r}$$

$$A_z = \sqrt{\frac{1}{8\mu E}} \left[(p_x L_y - p_y L_x) - (L_x p_y - L_y p_x)\right] - \sqrt{\frac{\mu}{2E}}\, \frac{x_3}{r} \qquad (5)$$

where E is the magnitude of the well-known Bohr energy level. A number of relations exist among these operators:

$$[L_x, L_y] = i\, L_z \qquad \text{and cyclically}$$

$$[A_x, L_x] = 0;\ [A_x, L_y] = iA_z \qquad \text{and cyclically}$$

$$[A_x, A_y] = iL_z \qquad \text{and cyclically}$$

$$A_x L_x + A_y L_y + A_z L_z = 0 = L_x A_x + L_y A_y + L_z A_z$$

$$L_x^2 + L_y^2 + L_z^2 + A_x^2 + A_y^2 + A_z^2 = \frac{\mu e^4}{2E} - 1 = n^2 - 1$$

(for a given energy level $E = E_n$) . (6)

If we define operators that differ from $\vec{L}$ and $\vec{A}$ only by a complex factor

$$L_1 = \frac{L_x}{i}\text{, etc.}$$

$$A_1 = \frac{A_x}{i}\text{, etc.} \qquad (7)$$

we see immediately that the equations in (5) now become

$$[L_i, L_j] = \sum_k \varepsilon_{ijk} L_k \qquad i,j = 1,2,3 .$$

$$[A_i, A_j] = \sum_k \varepsilon_{ijk} L_k$$

$$[L_i, A_j] = \sum_k \varepsilon_{ijk} A_k$$

$$F = L_1^2 + L_2^2 + L_3^2 + A_1^2 + A_2^2 + A_3^2 = \text{Constant}$$

$$G = L_1A_1 + L_2A_2 + L_3A_3 = 0 = A_1L_1 + A_2L_2 + A_3L_3 \qquad (8)$$

Comparison of Equations (7) and (2) shows that the quantum mechanical orbital angular momentum and Runge-Lenz-Pauli vector operators are just the infinitesimal generators of a four-dimensional rotation group O(4). Thus, the bound state wavefunctions of the hydrogen atom are basis functions for the irreducible representations of O(4).

It is well known that the Schroedinger equation can be solved in spherical polar as well as parabolic coordinates. This is also related to the O(4) symmetry. Defining new operators that are linear combinations of the six infinitesimal generators

$$j_{1i} = \frac{1}{2}(L_i + A_i)$$

$$j_{2i} = \frac{1}{2}(L_i - A_i) \qquad (9)$$

we see from Equations (6) and (7) that

$$[j_{1i}, j_{2j}] = 0 \quad \text{for any } i, j \qquad (10)$$

$$\left.\begin{aligned} [j_{1i}, j_{ij}] &= \sum_k \varepsilon_{ijk} j_{1k} \\ [j_{2i}, j_{2j}] &= \sum_k \varepsilon_{ijk} j_{2k} \end{aligned}\right\}$$

$$j_{1i} + j_{2i} = L_i \qquad (11)$$

$$\sum_i \{j_{1i}j_{1i} + j_{2i}j_{2i}\} = \frac{1}{2}\sum_i (L_i^{\,2} + A_i^{\,2})$$

or

$$\vec{j}_1 + \vec{j}_2 = \vec{L}$$

$$\vec{j}_1^{\,2} + \vec{j}_2^{\,2} = \frac{1}{2}(\vec{L}^2 + \vec{A}^2) \qquad (12)$$

Equations (10) and (11) show that $\vec{j}_1$ and $\vec{j}_2$ are infinitesimal generators of a three-dimensional rotation group, and since the two sets of operators commute, O(4) is a direct product of these two groups. The respective basis functions satisfy the eigenvalue equations

$$\vec{j}_1^{\,2}\,\phi_{j_1}^{m_1} = j_1(j_1 + 1)\,\phi_{j_1}^{m_1}$$

$$j_{1z}\,\phi_{j1}^{m_1} = m_1\,\phi_{j_1}^{m_1}$$

$$\vec{j}_2^{\,2}\,\phi_{j_2}^{m_2} = j_2(j_2+1)\,\phi_{j_2}^{m_2}$$

$$j_{2z}\,\phi_{j_2}^{m_2} = m_2\,\phi_{j_2}^{m_2}$$

$$\underset{\sim}{\mu}\,\phi_{j_1}^{m_1}\,\phi_{j_2}^{m_2} = E_n\,\phi_{j_1}^{m_1}\,\phi_{j_2}^{m_2} \qquad (13)$$

where $j_1\ m_1$ label the irreducible representations of the first rotation group and $j_2\ m_2$ of the second. The ψ_{nlm}, the solutions of $\underset{\sim}{H}$ in spherical polar coordinates, satisfy

$$H\,\psi_{nlm} = E_n\,\psi_{nlm}$$

$$\vec{L}^2\,\psi_{nlm} = l(l+1)\,\psi_{nlm}$$

$$L_z\,\psi_{nlm} = m\,\psi_{nlm} \qquad (14)$$

Equations (10) and (11) suggest a Clebsch-Gordan relationship between the basis functions of the irreducible representations of the direct product group ψ_{nlm} and the basis functions $\phi_{j_1}^{m_1} \phi_{j_2}^{m_2}$ of the irreducible representations of the groups that are factors of the direct product:

$$\psi_{nlm} = \sum_{m_1 m_2} \langle j_1 m_1 \; j_2 m_2 | lm \rangle \; \phi_{j_1}^{m_1} \phi_{j_2}^{m_2} \tag{15}$$

where the summation is restricted by the condition $m_1 + m_2 = m$, and <l> is a Clebsch-Gordan coefficient in the notation of *Physical Review*. Bargmann has shown that $\phi_{j_1}^{m_1} \phi_{j_2}^{m_2}$ is the wave function is parabolic coordinates and is a simultaneous eigenfunction of A_3 and L_3

$$A_3 \, \phi_{j_1}^{m_1} \phi_{j_2}^{m_2} = i(n_2 - n_1) \, \phi_{j_1}^{m_1} \phi_{j_2}^{m_2}$$

$$L_3 \, \phi_{j_1}^{m_1} \phi_{j_2}^{m_2} = - \, i \, m \, \phi_{j_1}^{m_1} \phi_{j_2}^{m_2} \tag{16}$$

n_1 and n_2 are the parabolic quantum numbers which are related to to the spherical quantum numbers n, l, and m.

$$m = n_1 + n_2 + |m| + 1$$

The other relations connecting the two sets of quantum numbers are

$$j_1 = \frac{1}{2} n(n-1) = j_2$$

$$m_1 = \frac{1}{2} (n_1 - n_2 + m)$$

$$m_2 = \frac{1}{2} (n_2 - n_1 + m) \tag{17}$$

In the language of Dirac quantum mechanics, the eigenfunctions ψ_{nlm} define a representation in which $\underset{\sim}{H}$, $\vec{L}^2$, and L_z are diagonal and $\phi_{j_1}^{m_1}\ \phi_{j_2}^{m_2}$ a representation in which $\underset{\sim}{H}$, A_z, and L_z are diagonal.

In the case of the O(3) group, the group of linear transformations that leave $x_1^2 + x_2^2 + x_3^2$ invariant, solutions of the three-dimensional Laplace equation are basis functions of its irreducible representations. This is because the spherical harmonics

$$Y_l^m(\theta,\phi)\ , \qquad \theta = \cos^{-1}\left(\frac{x_3}{r}\right), \qquad \phi = \tan^{-1}\left(\frac{x_2}{x_1}\right) \tag{18}$$

are the basis functions and the symmetry properties are not violated by multiplying Y_l^m with an arbitrary function of r. If, in particular, we multiply with r^l, then $r^l\ Y_l^m$ is a solution of Laplace equation. The basis functions for the irreducible representations of the four-dimensional rotation group O(4) are likewise solutions of a four-dimensional Laplace equation

$$\left(\frac{\partial^2}{\partial x_1^2} + \frac{\partial^2}{\partial x_2^2} + \frac{\partial^2}{\partial x_3^2} + \frac{\partial^2}{\partial x_4^2}\right)\Psi(r,\alpha,\theta,\phi) = 0 \tag{19}$$

This Laplace equation can be transformed into spherical polar coordinates, and the solutions are then essentially four-dimensional spherical harmonics.

Let

$$x_1 = r \cos\alpha$$

$$x_2 = r \sin\alpha \cos\theta$$

$$x_3 = r \sin\alpha \sin\theta \cos\phi$$

$$x_4 = r \sin\alpha \sin\theta \sin\phi$$

$$r = \sqrt{x_1^2 + x_2^2 + x_3^2 + x_4^2}$$

$$0 \le \alpha \le \pi$$

$$0 \le \theta \le \pi$$

$$0 \le \phi \le 2\pi \tag{20}$$

Then

$$\Psi(r,\alpha,\theta,\phi) = r^{n-1} f_n^{\ell}(\alpha)\, Y_{\ell}^{m}(\theta,\phi)$$

$$\left\{\frac{1}{r^3}\frac{\partial}{\partial r}\left(r^3\frac{\partial}{\partial r}\right) + \frac{1}{r^2}\left[\frac{1}{\sin^2\alpha}\frac{\partial}{\partial\alpha}\left(\sin^2\alpha\frac{\partial}{\partial\alpha}\right)\right]\right. \tag{21}$$

$$\left. + \frac{1}{r^2\sin^2\alpha}\left[\frac{1}{\sin\theta}\frac{\partial}{\partial\theta}\left(\sin\theta\frac{\partial}{\partial\theta}\right) + \frac{1}{\sin^2\theta}\frac{\partial^2}{\partial\phi^2}\right]\right\}\Psi(r,\alpha,\theta,\phi) = 0 \tag{22}$$

In the four-dimensional spherical harmonics $Y_1^m(\theta,\psi) f_n^{1}(\alpha)$, Y_1^{m} is the usual three-dimensional spherical harmonic and $f_n^1(\alpha)$ is

$$f_n^{\ell}(\alpha) = \left\{\frac{2^{21+1}\ n\ (n-\ell-1)!\ \ell!^2}{\pi(n+\ell)!}\right\}^{\frac{1}{2}} (\sin\alpha)^{\ell}\, C_{n-\ell-1}^{\ell+1}(\cos\alpha) \tag{23}$$

where $C_n^{\nu}(x)$ is the well-known Gegenbauer function. The f_n^{ℓ} satisfy the eigenvalue equation

$$\left(\frac{d^2}{d\alpha^2} + 2\cos\alpha\frac{d}{d\alpha} - \frac{1(1+1)}{\sin^2\alpha}\right) f_n^1(\alpha) = (1 - n^2)\, f_n^1(\alpha) \tag{24}$$

and are normalized

$$\int_0^{\pi} \{f_n^1(\alpha)\}^2 \sin^2\alpha\, d\alpha = 1 \tag{25}$$

Fock has shown that $\Psi(r,\alpha,\theta,\phi)$ are solutions of the Schroedinger equation in momentum space, n being the principal quantum number that determines the energy of the Bohr level.

WIGNER ECKART THEOREM

For all practical purposes, the Wigner-Eckart theorem can be expressed simply in the form of the equation

$$\left(\phi_{j'}^{m'},\ T_M^{(L)}\ \phi_j^m \right) = \langle jm\ LM|j'm'\rangle\ \langle j'||T^{(L)}||j\rangle \tag{26}$$

$T_M^{(L)}$ is the M component of an irreducible tensor of rank L with respect to SU(2); ϕ_j^m are basis functions of the (2j+1) dimensional irreducible representation of the latter; $\langle jm\ LM|j'm'\rangle$ is a Clebsch-Gordan coefficient; and $\langle j'||T^{(L)}||j\rangle$ is the so-called *reduced matrix element* which depends on the rank of the tensor L but is independent of m, m', and M. The m dependence of the matrix element is entirely contained in the Clebsch-Gordan coefficient.

A simple irreducible tensor of rank l = 1 can be constructed from the Pauli spin operators σ_x, σ_y, and σ_z with the three components labeled M - 1, O and -1

$$\sigma_1^{(1)} = -\frac{1}{\sqrt{2}}(\sigma_x + i\sigma_y),\quad \sigma_{-1}^{(1)} = \frac{1}{\sqrt{2}}(\sigma_x - i\sigma_y),$$

$$\sigma_0^{(1)} = \sigma_z \tag{27}$$

We illustrate the application of the Wigner-Eckart theorem by calculating matrix elements of these operators with the orthonormal basis functions ϕ_j^m belonging to $j = \frac{1}{2}$. We know from quantum mechanics

$$\sigma_x\ \phi_{\frac{1}{2}}^{\frac{1}{2}} = \phi_{\frac{1}{2}}^{-\frac{1}{2}} \qquad \sigma_y\ \phi_{\frac{1}{2}}^{\frac{1}{2}} = i\ \phi_{\frac{1}{2}}^{-\frac{1}{2}} \qquad \sigma_z\ \phi_{\frac{1}{2}}^{\frac{1}{2}} = \phi_{\frac{1}{2}}^{\frac{1}{2}}$$

$$\sigma_x\ \phi_{\frac{1}{2}}^{-\frac{1}{2}} = \phi_{\frac{1}{2}}^{\frac{1}{2}} \qquad \sigma_y\ \phi_{\frac{1}{2}}^{-\frac{1}{2}} = -i\ \phi_{\frac{1}{2}}^{\frac{1}{2}} \qquad \sigma_z\ \phi_{\frac{1}{2}}^{-\frac{1}{2}} = -\ \phi_{\frac{1}{2}}^{-\frac{1}{2}} \tag{28}$$

Since $\sigma_z = \sigma_o^{(1)}$, we readily see

$$(\phi_{\frac{1}{2}}^{\frac{1}{2}}, \sigma_{0}^{(1)} \phi_{\frac{1}{2}}^{\frac{1}{2}}) = (\phi_{\frac{1}{2}}^{\frac{1}{2}}, \phi_{\frac{1}{2}}^{\frac{1}{2}}) = 1$$

On the other hand, according to the theorem we should have

$$(\phi_{\frac{1}{2}}^{\frac{1}{2}}, \sigma_{0}^{(1)} \phi_{\frac{1}{2}}^{\frac{1}{2}}) = <\frac{1}{2}\frac{1}{2} 1\, 0 | \frac{1}{2}\frac{1}{2}> <\frac{1}{2}||\sigma^{(1)}||\frac{1}{2}> \qquad (29)$$

The Clebsch-Gordan coefficient $<\frac{1}{2}\frac{1}{2} 1\, 0 | \frac{1}{2}\frac{1}{2}>$ being $\frac{1}{\sqrt{3}}$, a comparison of the two equations shows that the reduced matrix element must be

$$<\frac{1}{2} || \sigma^{(1)} || \frac{1}{2}> = \sqrt{3} \qquad (30)$$

Now, for instance,

$$(\phi_{\frac{1}{2}}^{\frac{1}{2}}, \sigma_{1}^{(1)} \phi_{\frac{1}{2}}^{-\frac{1}{2}}) = <\frac{1}{2} -\frac{1}{2}\, 1\, 1 | \frac{1}{2}\frac{1}{2}> <\frac{1}{2} || \sigma^{(1)} || \frac{1}{2}>$$

$$= -\sqrt{\frac{2}{3}} \times \sqrt{3} = -\sqrt{2} \qquad (31)$$

using the numerical value of the Clebsch-Gordan coefficient $\left(= -\sqrt{\frac{2}{3}}\right)$. This, of course, can be seen more simply using Equation (27).

$$\sigma_{1}^{(1)} \phi_{\frac{1}{2}}^{-\frac{1}{2}} = -\frac{1}{\sqrt{2}} (\sigma_{x} + i\sigma_{y}) \phi_{\frac{1}{2}}^{-\frac{1}{2}} = -\frac{1}{\sqrt{2}} (1+1)\, \phi_{\frac{1}{2}}^{\frac{1}{2}} = -\sqrt{2}\, \phi_{\frac{1}{2}}^{\frac{1}{2}}$$

A slightly more involved application of the theorem is the evaluation of

$$(\phi_{\frac{1}{2}}^{\frac{1}{2}}, [\sigma_{1}^{(1)}, \sigma_{-1}^{(1)}]\, \phi_{\frac{1}{2}}^{\frac{1}{2}}),$$

that is, the expectation value of the commutator $[\sigma_{1}^{(1)}, \sigma_{-1}^{(1)}]$. We proceed as follows:

$$(\phi_{\frac{1}{2}}^{\frac{1}{2}} [\sigma_{1}^{(1)}, \sigma_{-1}^{(1)}]\, \phi_{\frac{1}{2}}^{\frac{1}{2}}) = (\phi_{\frac{1}{2}}^{\frac{1}{2}}, \sigma_{1}^{(1)} \sigma_{-1}^{(1)} \phi_{\frac{1}{2}}^{\frac{1}{2}})$$

$$- (\phi_{\frac{1}{2}}^{\frac{1}{2}}, \sigma_{-1}^{(1)} \sigma_{1}^{(1)} \phi_{\frac{1}{2}}^{\frac{1}{2}}) = \sum_{\tau}[(\phi_{\frac{1}{2}}^{\frac{1}{2}}, \sigma_{1}^{(1)} \phi_{\frac{1}{2}}^{\tau}) (\phi_{\frac{1}{2}}^{\tau}, \sigma_{-1}^{(1)} \phi_{\frac{1}{2}}^{\frac{1}{2}})$$

$$- (\phi_{\frac{1}{2}}^{\frac{1}{2}}, \sigma_{-1}^{(1)} \phi_{\frac{1}{2}}^{\tau}) (\phi_{\frac{1}{2}}^{\tau}, \sigma_{1}^{(1)} \phi_{\frac{1}{2}}^{\frac{1}{2}})]; \quad \tau = \pm \tfrac{1}{2} \qquad (32)$$

Using Wigner-Eckart theorem we have

$$(\phi_{\frac{1}{2}}^{\frac{1}{2}}, \sigma_{1}^{(1)} \phi_{\frac{1}{2}}^{\tau}) = < \frac{1}{2} \tau\, 1 1 \mid \frac{1}{2} \frac{1}{2} > < \frac{1}{2} || \sigma^{(1)} || \frac{1}{2} >$$

$$= \sqrt{3} < \frac{1}{2} \tau\, 1 1 \mid \frac{1}{2} \frac{1}{2} > \qquad (33)$$

Therefore, the expectation value reduces to

$$\{\sum_{\tau} < \frac{1}{2} \tau\, 11 \mid \frac{1}{2} \frac{1}{2} > < \frac{1}{2} \frac{1}{2}\, 1\, -1 \mid \frac{1}{2} \tau >$$

$$- < \frac{1}{2} \frac{1}{2}\, 1 1 \mid \frac{1}{2} \tau > < \frac{1}{2} \tau\, 1\, -1 \mid \frac{1}{2} \frac{1}{2} >\}$$

$$\times (< \frac{1}{2} || \sigma^{(1)} || \frac{1}{2} >)^{2} \qquad (34)$$

From tables of Clebsch-Gordan coefficients the required coefficients are known to be

$$< \frac{1}{2} \tau\, 11 \mid \frac{1}{2} \frac{1}{2} > = -\sqrt{\frac{2}{3}}\ \delta_{\tau,-\frac{1}{2}}$$

$$< \frac{1}{2} \frac{1}{2}\, 1\, -1 \mid \frac{1}{2} \tau > = \sqrt{\frac{2}{3}}\ \delta_{\tau,-\frac{1}{2}}$$

$$< \frac{1}{2} \frac{1}{2}\, 1 1 \mid \frac{1}{2} \tau > = 0 = < \frac{1}{2} \tau\, 1\, -1 \mid \frac{1}{2} \frac{1}{2} > \qquad (35)$$

A simple numerical substitution leads to the end result

$$(\phi^{\frac{1}{2}}_{\frac{1}{2}}\ [\sigma^{(1)}_{1},\ \sigma^{(1)}_{-1}]\ \phi^{\frac{1}{2}}_{\frac{1}{2}}) = -\sqrt{\frac{2}{3}} \times \sqrt{\frac{2}{3}} \times (\sqrt{3})^2$$

$$= -\ 2 \tag{36}$$

It is interesting to note that the tensor of rank 0 (scalar) is

$$-\ \sigma^{(1)}_{1}\ \sigma^{(1)}_{-1} - \sigma^{(1)}_{-1}\ \sigma^{(1)}_{1} + \sigma^{(1)}_{o}\ \sigma^{(1)}_{o} = 3$$

SYMMETRY OF THE ISOTROPIC HARMONIC OSCILLATOR

[SU(3) - Roots, Weights]

The isotropic harmonic oscillator in quantum mechanics is an excellent example of invariance under SU(3), the group of unitary unimodular transformations in three dimensions. In units of $\hbar = c = m = 1$ the nonrelativistic Hamiltonian can be written

$$\underset{\sim}{H} = \frac{1}{2}\,(p_i p_i + \omega^2\, x_i x_i) \tag{37}$$

where repeated indices are summed over. The five independent components of the symmetric tensor (operator)

$$A_{ij} = \frac{1}{2\omega}\,(p_i p_j + \omega^2\, x_i x_j) \qquad i,j = 1,2,3 \tag{38}$$

commute with H, as do the three components of the orbital angular momentum L_x, L_y, and L_z. SU(3) is an eight-parameter group (r=8) and has eight infinetesimal generators. The Lie algebra of SU(3) can conveniently be expressed in terms of linear combinations of these eight operators which are isomorphic with the infinitesimal generators:

$$H_1 = \frac{1}{2\sqrt{3}} L_z \qquad H_2 = \frac{1}{6} (A_{11} + A_{22} - 2A_{33})$$

$$E_1 = \frac{1}{2\sqrt{6}} (A_{11} - A_{22} + 2i\, A_{12}) \qquad E_{-1} = \frac{1}{2\sqrt{6}} (A_{11} - A_{22} - 2i\, A_{12})$$

$$E_2 = \frac{1}{4\sqrt{3}} (L_x + i\, L_y - 2A_{13} - 2i\, A_{23}) \qquad E_{-2} = \frac{1}{4\sqrt{3}} (L_x - iL_y - 2A_{13} + 2i\, A_{23})$$

$$E_3 = \frac{1}{4\sqrt{3}} (L_x + i\, L_y + 2A_{13} + 2i\, A_{23}) \qquad E_{-3} = \frac{1}{4\sqrt{3}} (L_x - iL_y + 2A_{13} - 2i\, A_{23})$$

$$[H_1, H_2] = 0$$

$$[H_1, E_1] = \sqrt{\frac{1}{3}}\, E_1, \quad [H_1, E_{-1}] = -\frac{1}{\sqrt{3}} E_{-1}, \quad [H_2, E_1] = 0 = [H_2, E_{-1}]$$

$$[H_1, E_2] = \frac{1}{2\sqrt{3}} E_2, \quad [H_1, E_{-2}] = -\frac{1}{2\sqrt{3}} E_{-2}, \quad [H_2, E_2] = \frac{1}{2} E_2, \quad [H_2, E_{-2}] = -\frac{1}{2} E_{-2}$$

$$[H_1, E_3] = \frac{1}{2\sqrt{3}} E_3, \quad [H_1, E_{-3}] = -\frac{1}{2\sqrt{3}} E_{-3}, \quad [H_2, E_3] = -\frac{1}{2} E_3 \quad [H_2, E_{-3}] = \frac{1}{2} E_{-3}$$

$$[E_1, E_{-1}] = \frac{1}{\sqrt{3}} H_1 \qquad [E_2, E_{-2}] = \frac{1}{2\sqrt{3}} H_1 + \frac{1}{2} H_2$$

$$[E_3, E_{-3}] = \frac{1}{2\sqrt{3}} H_1 - \frac{1}{2} H_2 \tag{39}$$

Writing the remaining commutators $[E_\alpha, E_\beta] = C^\gamma_{\alpha\beta} E_\gamma$ (no summation on γ) the structure constants turn out to be

$$C^{1}_{2,3} = C^{-1}_{-3,-2} = C^{2}_{1,-3} = C^{-3}_{-1,2} = C^{3}_{-2,1} = C^{-2}_{3,-1} = \frac{1}{\sqrt{6}} \quad (40)$$

and furthermore

$$[E_1, E_2] = 0 = [E_1, E_3] = [E_{-1}, E_{-2}] = [E_{-1}, E_{-3}]$$
$$= [E_2, E_{-3}] = [E_{-2}, E_3]$$

A straightforward calculation gives the Casimir invariant

$$\underset{\sim}{C} = H_1H_1 + H_2H_2 + E_1E_{-1} + E_{-1}E_1 + E_2E_{-2} + E_{-2}E_2 + E_3E_{-3} + E_{-3}E_3 \quad (41)$$

ROOT VECTORS, WIGHT VECTORS

The number of mutually commuting generators of this group, called the rank of the group ℓ, is 2 in this case, and these generators are H_1 and H_2, which explains the notational difference between these generators and the other six. The structure relations can be written compactly as

$$[H_i, H_j] = 0 \qquad i,j = 1,2\cdots\ell \quad (\text{here } \ell = 2)$$
$$[H_i, E_\alpha] = r_i(\alpha)E_\alpha \qquad \alpha = \pm 1, \pm 2, \pm 3 \cdots\cdot$$
$$[E_\alpha, E_{-\alpha}] = r_i(\alpha)H_i$$
$$[E_\alpha, E_\beta] = C^\gamma_{\alpha\beta} E_\gamma \qquad \alpha\neq\beta = \pm 1, \pm 2, \pm 3 \quad (42)$$

$r_i(\alpha)$ is considered as the i-th component of an ℓ-dimensional "root vector" $\vec{r}(\alpha)$. The different root vectors and their components are

$$\vec{r}(1) \rightarrow \left(\frac{1}{\sqrt{3}}, 0\right) \quad r_1(1) = \frac{1}{\sqrt{3}} \quad \vec{r}(-1) \rightarrow \left(-\frac{1}{\sqrt{3}}, 0\right)$$

$$r_2(1) = 0$$

$$\vec{r}(2) \rightarrow \left(\frac{1}{2\sqrt{3}}, \frac{1}{2}\right) \qquad \vec{r}(-2) \rightarrow \left(-\frac{1}{2\sqrt{3}}, -\frac{1}{2}\right)$$

$$\vec{r}(3) \rightarrow \left(\frac{1}{2\sqrt{3}}, -\frac{1}{2}\right) \qquad \vec{r}(-3) \rightarrow \left(-\frac{1}{2\sqrt{3}}, \frac{1}{2}\right) \tag{43}$$

These are represented graphically in Figure 1 in a Root Diagram with r_1 and r_2 as rectangular axes.

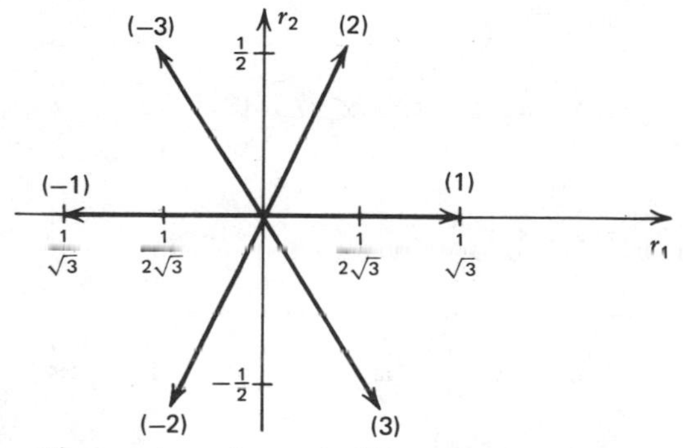

Figure 1. A root diagram.

The root vectors have the "ortho-normal" property

$$\sum_\alpha r_i(\alpha)\, r_j(\alpha) = \delta_{ij}$$

For instance, when i = 1 and j = 2 this equation becomes

$$\left(\frac{1}{\sqrt{3}} \times 0\right) + \left(-\frac{1}{\sqrt{3}} \times 0\right) + \left(\frac{1}{2\sqrt{3}} \times \frac{1}{2}\right) + \left(-\frac{1}{2\sqrt{3}} \times -\frac{1}{2}\right)$$

$$+ \left(\frac{1}{2\sqrt{3}} \times -\frac{1}{2}\right) + \left(-\frac{1}{2\sqrt{3}} \times \frac{1}{2}\right) = 0 \tag{44}$$

In the relation $[E_\alpha, E_\beta] = C^\gamma_{\alpha\beta} E_\gamma$ the structure constants $C^\gamma_{\alpha\beta}$ vanish whenever $\vec{r}(\alpha) + \vec{r}(\beta)$ is not a root vector. Thus

$$\vec{r}(2) + \vec{r}(-3) = \left(\frac{1}{2\sqrt{3}}, \frac{1}{2}\right) + \left(-\frac{1}{2\sqrt{3}}, \frac{1}{2}\right) = (0,1) \tag{45}$$

which is not a root vector, and hence $[E_2, E_{-3}] = 0$
The root diagram is thus a concise way of showing the structure of the Lie algebra.

The well-known solutions of the isotropic harmonic oscillator can be used as basis functions for the irreducible representations of the SU(3) group, and enumeration of the degenerate states gives us the dimensionality of the representations. Using the familiar nuclear shell model classification of states the central field function which is a solution of the Hamiltonian is written as

$$U_{nlm}(r,\theta,\phi) = F_{nl}(r)\, Y_l^m(\theta,\phi)$$

$$\frac{1}{2}(p_i p_i + \omega^2 x_i x_i)\, U_{nlm} = [2(n-1) + \ell + 3/2]\omega$$

$$(U_{n'l'm'}, U_{nlm}) = \delta_{nn'}\, \delta_{ll'}\, \delta_{mm'} \tag{46}$$

Energy $\frac{3}{2}\omega$ U_{100} (1)   (1s states)

$\frac{5}{2}\omega$ U_{111} U_{11-1} U_{110} (3)   (1p states)

$\frac{7}{2}\omega$ U_{200} $U_{1,2,2}$ $U_{1,2,1}$
$U_{1,2,0}$ $U_{1,2,-1}$ $U_{1,2,-2}$ (6) (2s, 1d states)

$\frac{9}{2}\omega$ $U_{2,1,1}$ $U_{2,1,-1}$ $U_{2,1,0}$
$U_{1,3,3}$ $U_{1,3,2}$ $U_{1,3,1}$ $U_{1,3,0}$
$U_{1,3,-1}$ $U_{1,3,-2}$ $U_{1,3,-3}$ (10)   (2p, 1f)

. . . . . .   (47)

We thus obtain 1, 3, 6, 10-dimensional representations. A straightforward calculation shows, for instance, that the three basis functions $U_{1,1,1}$, $U_{1,1,-1}$, $U_{1,1,0}$ are simultaneous eigenfunctions of the commuting generators H_1 and H_2, with the eigenvalues m_1 and m_2:

		m_1	m_2
$\underset{\sim}{H}_1 U_{1,1,1} = \frac{1}{2\sqrt{3}} U_{1,1,1}$	$\underset{\sim}{H}_2 U_{1,1,1} = \frac{1}{6} U_{1,1,1}$	$\frac{1}{2\sqrt{3}}$	$\frac{1}{6}$
$\underset{\sim}{H}_1 U_{1,1,-1} = -\frac{1}{2\sqrt{3}} U_{1,1,-1}$	$\underset{\sim}{H}_2 U_{1,1,-1} = \frac{1}{6} U_{1,1,-1}$	$-\frac{1}{2\sqrt{3}}$	$\frac{1}{6}$
$\underset{\sim}{H}_1 U_{1,1,0} = 0 \quad U_{1,1,0}$	$\underset{\sim}{H}_2 U_{1,1,0} = -\frac{2}{6} U_{1,1,0}$	0	$-\frac{2}{6}$

(48)

m_1 and m_2 are components of a vector in ℓ-dimensional space called a *weight vector*. The weight vectors are shown graphically in Figure 2 on a weight diagram for each representation. The explicit matrices, with the matrix elements of the generators calculated in the basis of these three eigenfunctions are given below.

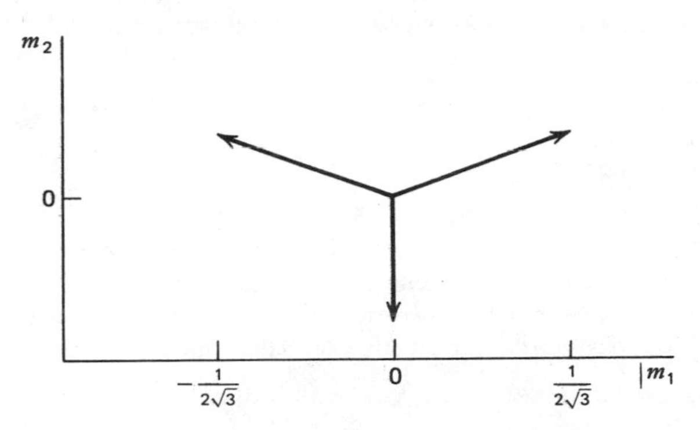

Figure 2. A weight diagram.

$$H_1 \rightarrow \frac{1}{2\sqrt{3}} \begin{pmatrix} 1 & 0 & 0 \\ 0 & -1 & 0 \\ 0 & 0 & 0 \end{pmatrix} \quad H_2 \rightarrow \frac{1}{6} \begin{pmatrix} 0 & 0 & 0 \\ 0 & 1 & 0 \\ 0 & 0 & -2 \end{pmatrix} \quad E_{-1} \rightarrow \frac{1}{\sqrt{6}} \begin{pmatrix} 0 & 0 & 0 \\ 1 & 0 & 0 \\ 0 & 0 & 0 \end{pmatrix}$$

$$E_1 \rightarrow \frac{1}{\sqrt{6}} \begin{pmatrix} 0 & 1 & 0 \\ 0 & 0 & 0 \\ 0 & 0 & 0 \end{pmatrix} \quad E_2 \rightarrow \frac{1}{\sqrt{6}} \begin{pmatrix} 0 & 0 & 1 \\ 0 & 0 & 0 \\ 0 & 0 & 0 \end{pmatrix} \quad E_{-2} \rightarrow \frac{1}{\sqrt{6}} \begin{pmatrix} 0 & 0 & 0 \\ 0 & 0 & 0 \\ 1 & 0 & 0 \end{pmatrix}$$

$$E_3 \to \frac{1}{\sqrt{6}}\begin{pmatrix} 0 & 0 & 0 \\ 0 & 0 & 0 \\ 0 & 1 & 0 \end{pmatrix} \quad E_{-3} \to \frac{1}{\sqrt{6}}\begin{pmatrix} 0 & 0 & 0 \\ 0 & 0 & 1 \\ 0 & 0 & 0 \end{pmatrix} \tag{49}$$

The role SU(3) symmetry plays in particle physics is discussed in a later chapter.

SPLITTING OF ATOMIC LEVELS

As far as symmetry is concerned, the energy levels of a free atom can be classified according to the irreducible representations of the rotation group. If we ignore spin, the degeneracy of the levels is $2l + 1$, where l represents angular momentum quantum number on the one hand and on the other it identifies the irreducible representation with the spherical harmonics Y_l^m as basis functions. We calculate and see how the levels of the atom are split when it is in a crystalline electric field of D_{3d} symmetry. The elements of $D_{3d} = D_3 \otimes i$ form a subgroup of the full rotation-reflection group, and hence the irreducible representations of the latter become reducible representations of the subgroup. The reduction theorem

$$a_\mu = \frac{1}{g}\sum_i g_i \chi_i^{(\mu)*} \chi_i \tag{50}$$

gives a_μ the number of times the μ-th irreducible representation of the subgroup is contained in the reducible representation in which elements of the i-th class have the compound character χ_i. g_i is the number of elements in class K_i that has the character $\chi_i^{(\mu)}$ in the μ-th representation. The character table of D_{3d} (Table I) specifies g_i, K_i, and χ_i.

The irreducible representations are classified as even (g) or odd (u) because of the presence of inversion (or the parity operation) as an element of the subgroup. In the pure rotation group D_3 the rotation angles that correspond to elements in classes K_1, K_2, and K_3 are, respectively, 0, $2\pi/3$ and $2\pi/2$, although the axes for the cyclic and dihedral rotations are different. The axes are immaterial to the calculation of *characters*, however, since these depend only on the angles according to the formula

$$\chi^{(\ell)}(\alpha) = \frac{\text{Sin } (\ell + \tfrac{1}{2})\alpha}{\text{Sin } (\alpha/2)} \qquad \alpha = \text{angle of rotation.} \tag{51}$$

TABLE I. CHARACTER TABLE OF D_{3d}

	$K_1(E)$ $g_1 = 1$	$K_2(C_3)$ $g_2 = 2$	$K_3(C_2)$ $g_3 = 3$	$K_4(i)$ $g_4 = 1$	$K_5(S_6 = i \times C_3)$ $g_5 = 2$	$K_6(\sigma_d = i \times C_2)$ $g_6 = 3$
A_{1g} $\mu=1$	1	1	1	1	1	1
A_{2g} $\mu=2$	1	1	-1	1	1	-1
E_g $\mu=3$	2	-1	0	2 $= \chi_4^{(3)}$	-1	0
A_{1u} $\mu=4$	1	1	1	-1	-1	-1
A_{2u} $\mu=5$	1	1	-1	-1	-1	1
E_{2u} $\mu=6$	2	-1	0	-2	1	0

If we apply this formula, the compound characters for the three classes of D_3 are

$$\chi^{(\ell)}(E) = \chi^{(\ell)}(\alpha)\Big|_{\alpha=0} = \frac{\sin(\ell+\frac{1}{2})\alpha}{\sin(\alpha/2)}\Big|_{\alpha=0} = 2\ell+1$$

$$\chi^{(\ell)}(C_3) = \chi^{(\ell)}(\alpha)\Big|_{\alpha=\frac{2\pi}{3}} = \begin{matrix} 1 & \ell = 0, 3, 6 \\ 0 & \ell = 1, 4, 7 \\ -1 & \ell = 2, 5, 8 \end{matrix}$$

$$\chi^{(\ell)}(C_2) = \chi^{\ell}(\alpha)\Big|_{\alpha=\pi} = (-1)^{\ell} \tag{51}$$

Because the spherical harmonics Y_1^m have parity $(-)^1$, the characters of the elements in the classes $K_5(S_6)$ and $K_6(\sigma_d)$ will be either ± the corresponding characters of elements in classes K_2 and K_3, since $S_6 = i \times C_3$ and $\sigma_d = i \times C_2$, the + sign going with even 1 and - sign with odd 1. Similarly, the characters of element i will be ± (21 + 1), since i = i x E. The complete compound character table of the elements of the subgroup D_{3d} of the rotation-reflection group is given in Table II for 1 = 0, 1, 2, 3, 4, 5 representations.

TABLE II. COMPOUND CHARACTER TABLE OF SUBGROUP D_{3d}

	E	$2C_3$	$3C_2$	i	$2S_6$	$3\sigma_d$
$\chi^{(\ell=6)}$	1	1	1	1	1	1
$\chi^{(1)}$	3	0	-1	-3	0	1
$\chi^{(2)}$	5	-1	1	5	-1	1
$\chi^{(3)}$	7	1	-1	-7	-1	1
$\chi^{(4)}$	9	0	1	9	0	1
$\chi^{(5)}$	11	-1	-1	-11	1	1

Now if we let 1 = 3, μ = 3, we then have

$$a_3 = \frac{1}{12}\{(1x2x7) + (2x-1x1) + (3x0x-1) + (1x2x-7)$$

$$+ (2x-1x-1) + (3x0x1)\}$$

$$= 0$$

and for μ = 6 we have

$$a_6 = \frac{1}{12}\{1x2x7) + (2x-1x1) + (3x0x-1) + (1x-2x-7)$$

$$+ \quad (2\times1\times-1) + (3\times0\times1)\}$$

$$= \quad 2 \tag{52}$$

Similar simple calculations give $a_1 = 0$, $a_2 = 0$, $a_3 = 0$, $a_4 = 1$, $a_5 = 2$, and $a_6 = 2$. Thus, the sevenfold degenerate level $l = 3$ of the free atom is split into one level of symmetry A_{1u}, two levels of symmetry A_{2u}, and two degenerate levels each of symmetry E_u.

$$E_3 \rightarrow A_{1u} + 2\,A_{2u} + 2\,E_u$$

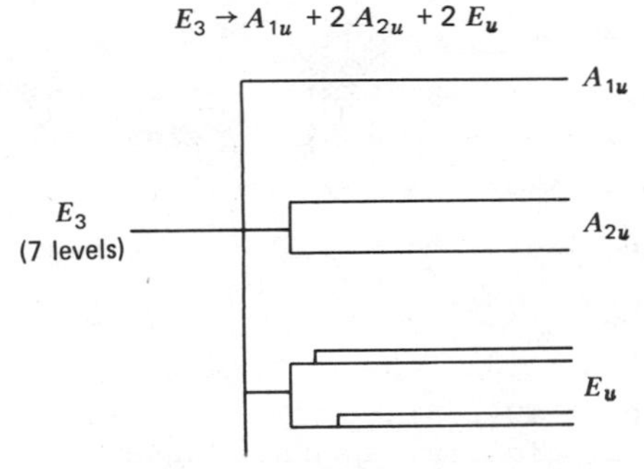

An important point to note is that the seven levels that are of odd parity ($(-)^3$) are split only into levels of odd symmetry with respect to D_{3d}. From the character tables by mere inspection we can calculate the splitting of the other levels

$$E_o \rightarrow A_{1g}$$

$$E_1 \rightarrow A_{2u} + E_u$$

$$E_2 \rightarrow A_{ig} + 2E_g$$

$$E_3 \rightarrow A_{1u} + 2A_{2u} + 2E_u$$

$$E_4 \rightarrow 2A_{1g} + A_{2g} + 3E_g$$

$$E_5 \rightarrow A_{1u} + 2A_{2u} + 4E_u$$

It is obvious that the work is a lot simpler in the case of crystal point groups which do not have a center of symmetry.

DOUBLE GROUPS

When the spin of the electron is taken into consideration, the splitting of levels is calculated according to the "double group" procedure introduced by Bethe (1929). Let us illustrate this for the dihedral group D_4 of order 8 which has three cyclic rotations (C_4, C_4^2, C_4^3) and two sets of dihedral rotations (D_x, D_y and D_1, D_2) which are easily visualized with the help of a square lamina. These eight elements of rotations can be obtained in a straightforward way from the Eulerian matrix for the three-dimensional rotation of a vector in space.

$$R(\alpha,\beta,\gamma) = \begin{pmatrix} \cos\alpha\cos\beta\cos\gamma - \sin\alpha\sin\gamma & -\sin\alpha\cos\beta\cos\gamma - \cos\alpha\sin\gamma & \sin\beta\cos\gamma \\ \cos\alpha\cos\beta\sin\gamma + \sin\alpha\cos\gamma & -\sin\alpha\cos\beta\sin\gamma + \cos\alpha\cos\gamma & \sin\beta\sin\gamma \\ -\cos\alpha\sin\beta & \sin\alpha\sin\beta & \cos\beta \end{pmatrix} \quad (53)$$

To obtain the two-valued spin representations we must write the elements of the subgroup of SU(2) into which the elements of the rotation group are homomorphically mapped. These are simply obtained by substitution of the angles α, β, γ in the matrix

$$\pm \begin{pmatrix} \cdot \cos(\beta/2)\, e^{\frac{i}{2}(\alpha+\gamma)} & \sin(\beta/2)\, e^{\frac{i}{2}(\gamma-\alpha)} \\ -\sin(\beta/2)\, e^{\frac{i}{2}(\alpha-\gamma)} & \cos(\beta/2)\, e^{-\frac{i}{2}(\alpha+\gamma)} \end{pmatrix} \quad (54)$$

with the convention that a prime is used in the notation of the element whenever the - sign of the matrix is taken. Thus when $\alpha = -\pi/2$, $\beta = \pi$, $\gamma = \pi/2$ we have

$$D_x = \begin{pmatrix} 0 & e^{\frac{\pi}{2}i} \\ -e^{-\frac{\pi i}{2}} & 0 \end{pmatrix} \qquad D'_x = \begin{pmatrix} 0 & -e^{\frac{\pi}{2}i} \\ e^{-\frac{\pi}{2}i} & 0 \end{pmatrix} \quad (55)$$

The elements can be written more elegantly, however, in terms of the eighth root of unity $\eta = \exp(\pi i/4)$ such that $\eta^8 = 1$.

$$E = \begin{pmatrix} \eta^8 & o \\ o & \eta^8 \end{pmatrix}: \quad C_4 = \begin{pmatrix} \eta & o \\ o & \eta^7 \end{pmatrix}: \quad C_4^2 = \begin{pmatrix} \eta^2 & o \\ o & \eta^6 \end{pmatrix}: \quad C_4^3 = \begin{pmatrix} \eta^3 & o \\ o & \eta^5 \end{pmatrix}$$

$$R \equiv E' = \begin{pmatrix} \eta^4 & o \\ o & \eta^4 \end{pmatrix}: \quad C_4' = C_4R = \begin{pmatrix} \eta^5 & o \\ o & \eta^3 \end{pmatrix}: \quad C_4^{2'} = \begin{pmatrix} \eta^6 & o \\ o & \eta^2 \end{pmatrix}: \quad C_4^{3'} = \begin{pmatrix} \eta^7 & o \\ o & \eta \end{pmatrix}$$

$$D_x = \begin{pmatrix} o & \eta^2 \\ \eta^2 & o \end{pmatrix}: \quad D_y = \begin{pmatrix} o & \eta^8 \\ \eta^4 & o \end{pmatrix}: \quad C_1 \doteq \begin{pmatrix} o & \eta^5 \\ \eta^7 & o \end{pmatrix}: \quad D_2 = \begin{pmatrix} o & \eta^7 \\ \eta^5 & o \end{pmatrix}$$

$$D_x' = \begin{pmatrix} o & \eta^6 \\ \eta^6 & o \end{pmatrix}: \quad D_g' = \begin{pmatrix} o & \eta^4 \\ \eta^8 & o \end{pmatrix}: \quad D_1' = \begin{pmatrix} o & \eta \\ \eta^3 & o \end{pmatrix}: \quad D_2' = \begin{pmatrix} o & \eta^3 \\ \eta & o \end{pmatrix} \qquad (56)$$

This "double group" of order 16 satisfies the multiplication table on the Table III. There are seven classes:

$K_1 \to E \qquad g_1 = 1$

$K_2 \to R \qquad g_2 = 1$

$K_3 \to C_4^{\,2}, C_4^{\,2'} \qquad g_3 = 2$

$K_4 \to C_4, C_4^{\,3'} \qquad g_4 = 2$

$K_5 \to C_4^{\,3}, C_{4'} \qquad g_5 = 2$

$K_6 \to D_x, D_y, D_x', D_y' \qquad g_6 = 4$

$K_7 \to D_1, D_1', D_2, D_2' \qquad g_7 = 4$

The group has four one-dimensional and three two-dimensional representations in accordance with the relation

$$1^2 + 1^2 + 1^2 + 1^2 + 2^2 + 2^2 + 2^2 = 16 \ .$$

The characters can be calculated applying the class formula (22) of Chapter Three:

$$g_i g_j \chi_i^{(\mu)} \chi_j^{(\mu)} = n_\mu \sum_\ell C_{ijl} \chi_\ell^{(\mu)} \qquad (57)$$

If now we take the six degenerate levels of the free atom corresponding to $j = 5/2$, the compound characters corresponding to the $(2j+1)$ dimensional representation can be evaluated according to

$$\frac{\sin (j + \frac{1}{2})\alpha}{\sin (\alpha/2)}$$

where α is 0 for elements in class K_1 and assumes the values π, $\pi/2$, $3\pi/2$, π and π, respectively, for elements in classes K_3, K_4, K_5, K_6 and K_7. These compound characters are

TABLE III. MULTIPLICATION TABLE FOR THE DOUBLE GROUP D'_4

	E	C_4	C_4^2	C_4^3	D_x	D_y	D_1	D_2	R	C'_4	$C_4^{2'}$	$C_4^{3'}$	D'_x	D'_y	D'_1	D'_2
E	E	C_4	C_4^2	C_4^3	D_x	D_y	D_1	D_2	R	C'_4	$C_4^{2'}$	$C_4^{3'}$	D'_x	D'_y	D'_1	D'_2
C_4	C_4	C_4^2	C_4^3	R	D'_2	D'_1	D'_x	D_y	C'_4	$C_4^{2'}$	$C_4^{3'}$	E	D_2	D_1	D_x	D'_y
C_4^2	C_4^2	C_4^3	R	C'_4	D'_y	D_x	D_2	D'_1	$C_4^{2'}$	$C_4^{3'}$	E	C_4	D_y	D'_x	D'_2	D_1
C_4^3	C_4^3	R	C'_4	$C_4^{2'}$	D_1	D'_2	D_y	D_x	$C_4^{3'}$	E	C_4	C_4^2	D'_1	D_2	D'_y	D'_x
D_x	D_x	D'_1	D_y	D_2	R	$C_4^{2'}$	C_4	$C_4^{3'}$	D'_x	D_1	D'_y	D'_2	E	C_4^2	C'_4	C_4^3
D_y	D_y	D_2	D'_x	D_1	C_4^2	R	$C_4^{3'}$	C'_4	D'_y	D'_2	D_x	D'_1	$C_4^{2'}$	E	C_4^3	C_4
D_1	D_1	D'_y	D'_2	D_x	$C_4^{3'}$	C_4	R	C_4^2	D'_1	D_y	D_2	D'_x	C_4^3	C'_4	E	$C_4^{2'}$
D_2	D_2	D'_x	D_1	D'_y	C_4	C_4^3	$C_4^{2'}$	R	D'_2	D_x	D'_1	D_y	C'_4	$C_4^{3'}$	C_4^2	E
R	R	C'_4	$C_4^{2'}$	$C_4^{3'}$	D'_x	D'_y	D'_1	D'_2	E	C_4	C_4^2	C_4^3	D_x	D_y	D_1	D_2

TABLE III (Continued)

	E	C_4	C_4^2	C_4^3	D_x	D_y	D_1	D_2	R	C'_4	$C_4^{2'}$	$C_4^{3'}$	D'_x	D'_y	D'_1	D'_2
C'_4	C'_4	$C_4^{2'}$	$C_4^{3'}$	E	D_2	D_1	D_x	D'_y	C_4	C_4^2	C_4^3	R	D'_2	D'_1	D'_x	D_y
$C_4^{2'}$	$C_4^{2'}$	$C_4^{3'}$	E	C_4	D_y	D'_x	D'_2	D_1	C_4^2	C_4^3	R	C'_4	D'_y	D_x	D_2	D'_1
$C_4^{3'}$	$C_4^{3'}$	E	C_4	C_4^2	D'_1	D_2	D'_y	D'_x	C_4^3	R	C'_4	$C_4^{2'}$	D_1	D'_2	D_y	D_x
D'_x	D'_x	D_1	D'_y	D'_2	E	C_4^2	C'_4	C_4^3	D_x	D'_1	D_y	D_2	R	$C_4^{2'}$	C_4	$C_4^{3'}$
D'_y	D'_y	D'_2	D_x	D'_1	$C_4^{2'}$	E	C_4^3	C_4	D_y	D_2	D'_x	D_1	C_4^2	R	$C_4^{3'}$	C'_4
D'_1	D'_1	D_y	D_2	D'_x	C_4^3	C'_4	E	$C_4^{2'}$	D_1	D'_y	D'_2	D_x	$C_4^{3'}$	C_4	R	C_4^2
D'_2	D'_2	D_x	D'_1	D_y	C'_4	$C_4^{3'}$	C_4^2	E	D_2	D'_x	D_1	D'_y	C_4	C_4^3	$C_4^{2'}$	R

K_1	K_2	K_3	K_4	K_5	K_6	K_7	
6	-6	0	$-\sqrt{2}$	$\sqrt{2}$	0	0	(58)

Comparison with the character table of D'_4 (Table IV) shows that this atomic level $E_{5/2}$ is split into a twofold degenerate level of symmetry E'_1 and two twofold degenerate levels each of symmetry E'_2

$$E_{5/2} \rightarrow E'_1 + 2E'_2 \ . \tag{59}$$

Thus two-valued representations are of importance for half-integral j values of the free atom, and these levels are split only into levels belonging to the two valued representations of D'_4.

TABLE IV. CHARACTER TABLE OF D'_4

	K_1	K_2	K_3	K_4	K_5	K_6	K_7
A_1	1	1	1	1	1	1	1
A_2	1	1	1	1	1	-1	-1
B_1	1	1	1	-1	-1	1	-1
B_2	1	1	1	-1	-1	-1	1
E	2	2	-2	0	0	0	0
E'_1	2	-2	0	$\sqrt{2}$	$-\sqrt{2}$	0	0
E'_2	2	-2	0	$-\sqrt{2}$	$\sqrt{2}$	0	0

E'_1 and E'_2 are the double-valued representations in which E and R have opposite characters.

CHAPTER SIX

APPLICATIONS IN NUCLEAR PHYSICS

L-S COUPLING SCHEME

When a many-nucleon central-field Hamiltonian does not have spin-dependent forces, its eigenfunction follows the L-S coupling scheme in general. $\vec{L}$ is the vector sum of the orbital angular momentum operators of all particles

$$\vec{L} = \sum_i \vec{\ell}_i \tag{1}$$

and $\vec{S}$ is likewise the sum of the spin vector operators

$$\vec{S} = \sum_i \vec{S}_i \tag{2}$$

The total wave function is a simultaneous eigenfunction of $\vec{L}^2$, L_z, $\vec{S}^2$, and S_z and the respective quantum numbers are usually denoted by L, M_L, S, M_S. The angular momentum eigenfunction of $\vec{L}^2$ is built up of single particle eigenfunctions of $\vec{\ell}_i^2$ applying the well-known Clebsch-Gordan theorem, and a similar procedure gives the eigenfunction of $\vec{S}^2$.

According to the Pauli principle, the many-nucleon wavefunction has to be antisymmetric with respect to the interchange of all the coordinates of any two particles. This is because the nucleons are spin ½ particles obeying Fermi-Dirac statistics and the Hamiltonian is symmetrical in the coordinates of all particles. The Hamiltonian will be invariant to permutations and thus belongs to the permutation group. Its eigenfunction should, therefore, be a basis function for an irreducible representation of the permutation group. We will construct a few wavefunctions following these symmetry requirements.

Let us first take the functions describing just the spin states of a three particle system, each of spin ½. We use the following notation for the spin up (↑) and spin down (↓) states of one particle

$$(\uparrow) = \chi_{\frac{1}{2}}^{\frac{1}{2}}(1) \equiv \alpha(1) \qquad (\downarrow)\ \chi_{\frac{1}{2}}^{-\frac{1}{2}}(1) \equiv \beta(1) \tag{3}$$

the number in parenthesis signifying that the state refers to particle one. We have

$$\begin{aligned} \vec{S} &- \vec{S}_1 + \vec{S}_2 + \vec{S}_3 \\ &\equiv \vec{S}_{12} + \vec{S}_3 \end{aligned} \tag{4}$$

$\chi_S^{M_S}(1,2,3)$ is the three-particle spin-wave function that is a simultaneous eigenfunction of $\vec{S}^2$ with eigenvalue $S(S+1)$ and S_z with eigenvalue M_S. The Clebsch-Gordan theorem gives the following combinations for

S = 3/2

$$\chi_{3/2}^{3/2}(1,2,3) \equiv \chi_{3/2}^{1/2} = \alpha(1)\ \alpha(2)\ \alpha(3)$$

$$\chi_{3/2}^{1/2} = \frac{1}{\sqrt{3}}\{\alpha(1)\alpha(2)\beta(3)+\alpha(2)\alpha(3)\beta(1)+\alpha(3)\alpha(1)\beta(2)\}$$

$$\chi_{3/2}^{-1/2} = \frac{1}{\sqrt{3}}\{\alpha(1)\beta(2)\beta(3)+\alpha(2)\beta(3)\beta(1)+\alpha(3)\beta(1)\beta(2)\}$$

$$\chi_{3/2}^{-3/2} = \beta(1)\beta(2)\beta(3) \tag{5}$$

(all totally symmetric)

S = 1/2

$$\chi_{1/2}^{1/2} = \frac{1}{\sqrt{6}}\{2\alpha(1)\alpha(2)\beta(3)-\alpha(2)\alpha(3)\beta(1)-\alpha(3)\alpha(1)\beta(2)\} \equiv \phi_3 \tag{6}$$

$$\chi^{-1/2}_{1/2} = \frac{1}{\sqrt{6}} \{-2\beta(1)\beta(2)\alpha(3)+\beta(2)\beta(3)\alpha(1)+\beta(3)\beta(1)\alpha(2)\}$$

(here $S_{12} = 1$)

$$\chi^{1/2}_{1/2} = \frac{1}{\sqrt{2}} \{\alpha(1)\ \beta(2)\ \alpha(3) - \alpha(2)\ \beta(1)\ \alpha(3)\} \equiv \phi_4$$

$$\chi^{-1/2}_{1/2} = \frac{1}{\sqrt{2}} \{\alpha(1)\ \beta(2)\ \beta(3) - \alpha(2)\ \beta(1)\ \beta(3)\} \tag{7}$$

(here $S_{12} = 0$)

We will generate the two-dimensional representation of S_3 with ϕ_3 and ϕ_4 as basis functions. S_3 is, as usual, the permutation group on three objects.

$$\underset{\sim}{C}\ \phi_3 \equiv (12)\ \phi_3 = \phi_3 \qquad \underset{\sim}{C}\ \phi_4 = -\ \phi_4$$

$$\text{Therefore, } D(\underset{\sim}{C}) = \begin{pmatrix} 1 & 0 \\ 0 & -1 \end{pmatrix} \qquad \chi = 0 \tag{8}$$

Straightforward calculations show that the matrices representing the other permutations of the group S_3 are

$$D(\underset{\sim}{E}) = \begin{pmatrix} 1 & 0 \\ 0 & 1 \end{pmatrix} \quad \chi = 2 \qquad D(A) = \begin{pmatrix} -\frac{1}{\sqrt{2}} & \frac{\sqrt{3}}{2} \\ \frac{\sqrt{3}}{2} & \frac{1}{2} \end{pmatrix} \quad \chi = 0$$

$$D(B) = \begin{pmatrix} -\frac{1}{2} & -\frac{\sqrt{3}}{2} \\ -\frac{\sqrt{3}}{2} & \frac{1}{2} \end{pmatrix} \quad \chi = 0 \qquad D(D) = \begin{pmatrix} -\frac{1}{2} & \frac{\sqrt{3}}{2} \\ -\frac{\sqrt{3}}{2} & -\frac{i}{2} \end{pmatrix} \quad \chi = -1$$

$$D(F) = \begin{pmatrix} -\frac{1}{2} & -\frac{\sqrt{3}}{2} \\ \frac{\sqrt{3}}{2} & -\frac{1}{2} \end{pmatrix}$$

$$\chi = -1 \tag{9}$$

TWO DIMENSIONAL REPRESENTATIONS OF S_3, BASIS FUNCTIONS

The characters χ prove that the basis set ϕ_3, ϕ_4 generates the irreducible representation of S_3. As Hamermesh (1962) has shown, the basis functions can be generated with the help of projection operators associated with Young diagrams. There are two standard Young tableaux for this two-dimensional representation:

[1] [2] [1] [3]
[3] [2]

and the projection operators pertaining to these diagrams are

$$\underset{\sim}{P}_1 = [E - (13)][E + (12)] = \{E + (12) - (13) - (123)\}$$

$$\equiv (E + C - B - D)$$

$$P_2 = [E - (12)][E + (13)] = \{E - (12) + (13) - (132)\}$$

$$= (E - C + B - F) \tag{10}$$

Operating on $\alpha(1)\alpha(2)\beta(3)$, these projection operators generate the basis functions

$$g_3 \equiv \underset{\sim}{P}_1\, \alpha(1)\alpha(2)\beta(3) = 6\,[\alpha(1)\alpha(2)\beta(3) - \alpha(2)\alpha(3)\beta(1)]$$

$$g_4 \equiv \underset{\sim}{P}_2\, \alpha(1)\alpha(2)\beta(3) = 3\,[\alpha(2)\alpha(3)\beta(1) - \alpha(1)\alpha(3)\beta(2)]$$

$$\tag{11}$$

Simple calculation yields the following representation of S_3:

$$D(E) = \begin{pmatrix} 1 & 0 \\ 0 & 1 \end{pmatrix} \quad D(A) = \begin{pmatrix} 0 & -\frac{1}{2} \\ -2 & 0 \end{pmatrix} \quad D(B) = \begin{pmatrix} -1 & \frac{1}{2} \\ 0 & 1 \end{pmatrix}$$

$$\chi = 2 \qquad\qquad \chi = 0 \qquad\qquad \chi = 0$$

$$D(C) = \begin{pmatrix} 1 & 0 \\ 2 & -1 \end{pmatrix} \quad D(D) = \begin{pmatrix} 0 & -\frac{1}{2} \\ 2 & -1 \end{pmatrix} \quad D(F) = \begin{pmatrix} -1 & \frac{1}{2} \\ -2 & 0 \end{pmatrix}$$

$$\chi = 0 \qquad\qquad \chi = -1 \qquad\qquad \chi = -1 \tag{12}$$

Comparison of the characters in the representations (9) and (12) shows that the two are irreducible and equivalent. Φ_3 can be related to the Young diagram

[1] [2]
[3]

and ϕ_4 to the (self) Conjugate diagram

[1] [3]
[2]

Consider now the orbital wavefunctions. If we denote ϕ_a, ϕ_b, ϕ_c, respectively, the spherical harmonics Y_1^1, Y_1^0, Y_1^{-1} multiplied by appropriate radial functions, the two-particle states D, P, S made up of two p nucleons are given, according to the Clebsch-Gordan theorem, by

D State

$$\psi_{L=2}^{M_L=2}(1,2) \equiv \psi_D^2 = \phi_a(1)\ \phi_a(2)$$

$$\psi_D^1 = \frac{1}{\sqrt{2}}\{\phi_a(1)\ \phi_b(2) + \phi_a(2)\ \phi_b(1)\}$$

$$\psi_D^0 = \frac{1}{\sqrt{6}}\{\phi_a(1)\ \phi_c(2) + \phi_a(2)\ \phi_c(1) + 2\phi_b(1)\phi_b(2)\}$$

$$\psi_D^{-1} = \frac{1}{\sqrt{2}}\{\phi_b(1)\ \phi_c(2) + \phi_b(2)\ \phi_c(1)\}$$

$$\psi_D^{-2} = \phi_c(1)\ \phi_c(2) \qquad \text{(symmetric)} \tag{13}$$

P AND D STATE WAVEFUNCTIONS OF APPROPRIATE SYMMETRY

P State

$$\left.\begin{aligned}\psi_P^1 &= \frac{1}{\sqrt{2}}\{\phi_a(1)\ \phi_b(2) - \phi_a(2)\ \phi_b(1)\} \\ \psi_P^0 &= \frac{1}{\sqrt{2}}\{\phi_a(1)\ \phi_c(2) - \phi_a(2)\ \phi_c(1)\} \\ \psi_P^{-1} &= \frac{1}{\sqrt{2}}\{\phi_b(1)\ \phi_c(2) - \phi_b(2)\ \phi_c(1)\}\end{aligned}\right\}\ \text{(antisymmetric)}$$

S State

$$\psi_S^0 = \frac{1}{\sqrt{3}}[\phi_a(1)\ \phi_c(2) + \phi_a(2)\ \phi_c(1) - \phi_b(1)\ \phi_b(2)]$$

(symmetric)

These are to be combined with $\phi_a(3)$, $\phi_b(3)$ and $\phi_c(3)$ in the three-particle system, where each single-particle state is assumed to be a p state. As an example, the three-particle state functions that are simulatneous eigenfunctions of $\vec{L}^2$ and L_z with eigenvalues $L = 2$, $M_L = 0$ are

$$\begin{aligned}\Psi_3 = \frac{1}{\sqrt{12}}\{&\phi_a(1)\phi_b(2)\phi_c(3) + \phi_b(1)\phi_c(2)\phi_a(3) \\ &- \phi_b(1)\phi_a(2)\phi_c(3) - \phi_c(1)\phi_b(2)\phi_a(3) \\ &+ 2\phi_a(1)\phi_c(2)\phi_b(3) - 2\psi_c(1)\phi_a(2)\phi_b(3)\}\end{aligned}$$

$$\begin{aligned}\Psi_4 = \frac{1}{\sqrt{12}}\{&\phi_a(1)\phi_b(2)\phi_c(3) + \phi_b(1)\phi_a(2)\phi_c(3) \\ &- \phi_c(1)\phi_b(2)\phi_a(3) - \phi_b(1)\phi_c(2)\phi_a(3)\}\end{aligned} \qquad (14)$$

These generate the two-dimensional irreducible representation of S_3:

$$D(E) = \begin{pmatrix} 1 & 0 \\ 0 & 1 \end{pmatrix} \quad D(C) = \begin{pmatrix} -1 & 0 \\ 0 & 1 \end{pmatrix} \quad D(A) = \begin{pmatrix} \frac{1}{2} & \frac{\sqrt{3}}{2} \\ \frac{\sqrt{3}}{2} & -\frac{1}{2} \end{pmatrix}$$

$$\chi = 2$$

$$D(B) = \begin{pmatrix} \frac{1}{2} & -\frac{\sqrt{3}}{2} \\ -\frac{\sqrt{3}}{2} & -\frac{1}{2} \end{pmatrix} \qquad D(D) = \begin{pmatrix} -\frac{1}{2} & -\frac{\sqrt{3}}{2} \\ -\frac{\sqrt{3}}{2} & -\frac{1}{2} \end{pmatrix}$$

$$D(F) = \begin{pmatrix} -\frac{1}{2} & \frac{\sqrt{3}}{2} \\ -\frac{\sqrt{3}}{2} & -\frac{1}{2} \end{pmatrix} \tag{15}$$

According to quantum mechanical L-S coupling scheme, the normalized three-particle wave function of the ^2P state with quantum numbers $L = 1$, $M_L = 0$, $S = \frac{1}{2}$, $M_S = \frac{1}{2}$ is

$$[\sqrt{\frac{4}{18}}\ \{\frac{1}{\sqrt{2}}\ \alpha(1)\beta(2)\alpha(3) - \frac{1}{\sqrt{2}}\ \alpha(2)\beta(1)\alpha(3)\}\{\frac{1}{\sqrt{3}}\ \phi_c(1)\phi_a(2)\phi_b(3)$$

$$+ \frac{1}{\sqrt{3}}\ \phi_c(2)\phi_a(1)\phi_b(3) - \frac{1}{\sqrt{3}}\ \phi_b(1)\phi_b(2)\phi_b(3)\}]$$

$$+ \sqrt{\frac{9}{18}}\ [\{-\sqrt{\frac{2}{3}}\ \alpha(1)\beta(3)\alpha(2) + \frac{1}{\sqrt{6}}\ \alpha(1)\beta(2)\alpha(3)$$

$$+ \frac{1}{\sqrt{6}}\ \alpha(2)\beta(1)\alpha(3)\}\ \{\frac{1}{2}\ \phi_a(1)\phi_b(2)\phi_c(3) - \frac{1}{2}\ \phi_a(2)\phi_b(1)\phi_c(3)$$

$$+ \frac{1}{2}\ \phi_a(3)\phi_b(2)\phi_c(1) - \frac{1}{2}\ \phi_a(3)\phi_b(1)\phi_c(2)\}]$$

$$+ \sqrt{\frac{5}{18}}\ [\{-\ \frac{1}{\sqrt{2}}\ \alpha(1)\beta(2)\alpha(3) + \frac{1}{\sqrt{2}}\ \alpha(2)\beta(1)\alpha(3)\}$$

$$\times\ \{\sqrt{\frac{3}{20}}\ \phi_b(1)\phi_c(2)\phi_a(3) + \sqrt{\frac{3}{20}}\ \phi_b(2)\phi_c(3)\phi_a(1)$$

$$+\sqrt{\frac{3}{20}}\ \phi_b(2)\phi_c(1)\phi_a(3) +\sqrt{\frac{3}{20}}\ \phi_b(1)\phi_c(3)\phi_a(2)$$

$$- \frac{1}{\sqrt{15}}\ \phi_b(3)\phi_c(2)\phi_a(1) - \frac{1}{\sqrt{15}}\ \phi_b(3)\phi_c(1)\phi_a(2)$$

$$-\sqrt{\frac{4}{15}}\ \phi_b(1)\phi_b(2)\phi_b(3)\}] \qquad (16)$$

This satisfies the Pauli principle requirement of antisymmetry in the interchange of all coordinates (space and spin) of any two particles. We have noticed earlier that the orbital functions and spin functions separately belong to the irreducible representations of S_3. In the right linear combination the orbital and spin functions are associated with conjugate Young diagrams. Using the notation

$$\phi_a(1)\ \alpha(1) \equiv \phi_a^{+}(1) \qquad\qquad \phi_a(1)\ \beta(1) \equiv \phi_a^{-}(1)$$

if we add and subtract

$$\phi_b^{-}(1)\ \phi_a^{+}(2)\ \phi_c^{+}(3) + \phi_a^{-}(1)\ \phi_a^{+}(3)\ \phi_c^{+}(2)$$

the above function in Equation (16) can be rewritten, as DeShalit and Talmi (1963) have shown in their book:

$$\frac{1}{\sqrt{12}} x\{[\alpha(1)\beta(2)\alpha(3) - \beta(1)\alpha(2)\alpha(3)][\phi_a(1)\phi_c(2)\phi_b(3)$$

$$+ \phi_c(1)\phi_b(2)\phi_a(3) - \phi_b(1)\phi_a(2)\phi_c(3)$$

$$- \phi_b(1)\phi_c(2)\phi_a(3)]$$

$$- [\alpha(1)\alpha(2)\beta(3) - \beta(1)\alpha(2)\alpha(3)][\phi_a(1)\phi_b(2)\phi_c(3)$$

$$+ \phi_c(1)\phi_b(2)\phi_a(3) - \phi_b(1)\phi_a(2)\phi_c(3)$$

$$- \phi_b(1)\phi_c(2)\phi_a(3)]\} \qquad (17)$$

In terms of projection operators associated with Young diagrams this can be shown as

$$\frac{1}{\sqrt{12}}\left\{\begin{array}{ll}[1] & [3]\\ {[2]} & \end{array}\ \alpha(1)\beta(2)\alpha(3)\quad \begin{array}{ll}[1] & [2]\\ {[3]} & \end{array}\ \phi_a(1)\phi_c(2)\phi_b(3)\right\}$$

$$-\frac{1}{\sqrt{12}}\left\{\begin{array}{ll}[1] & [2]\\ {[3]} & \end{array}\ \alpha(1)\alpha(2)\beta(3)\quad \begin{array}{ll}[1] & [3]\\ {[2]} & \end{array}\ \phi_a(1)\phi_b(2)\phi_c(3)\right\} \qquad (18)$$

As another example, let us take the L-S coupled wave function for the state $L = 2$, $M_L = 0$, $S = \frac{1}{2}$, $M_S = \frac{1}{2}$. From Equations (5), (6), (7), and (14) we obtain the normalized wave function quantum mechanically:

$$\left[-\frac{1}{\sqrt{12}}\alpha(1)\beta(2)\alpha(3) - \frac{1}{\sqrt{12}}\beta(1)\alpha(2)\alpha(3) + \frac{1}{\sqrt{3}}\alpha(1)\alpha(2)\beta(3)\right]$$

$$\times\left[\frac{1}{\sqrt{12}}\phi_a(1)\phi_b(2)\phi_c(3) + \frac{1}{\sqrt{12}}\phi_b(1)\phi_c(2)\phi_a(3)\right.$$

$$-\frac{1}{\sqrt{12}}\phi_b(1)\phi_a(2)\phi_c(3) - \frac{1}{\sqrt{3}}\phi_a(1)\phi_c(2)\phi_b(3)$$

$$\left.-\frac{1}{\sqrt{3}}\phi_c(1)\phi_a(2)\phi_b(3)\right] + \left[\frac{1}{2}\phi_a(1)\phi_b(2)\phi_c(3)\right.$$

$$+\frac{1}{2}\phi_b(1)\phi_a(2)\phi_c(3) - \frac{1}{2}\phi_c(1)\phi_b(2)\phi_a(3)$$

$$\left.-\frac{1}{2}\phi_b(1)\phi_c(2)\phi_a(3)\right] \times \left[\frac{1}{2}\beta(1)\alpha(2)\alpha(3) - \frac{1}{2}\alpha(1)\beta(2)\alpha(3)\right] \equiv$$

$$\chi_s\,\Psi_o + \chi_s'\,\Psi_o' \qquad (16)$$

In each of the two terms we have a product of a space function and a spin function of the three-particle state. It is easy to show that in Ψ_o if we let

$$a = b = \alpha, \qquad c = \beta$$

we obtain χ'_S apart from a numerical factor. Similarly, if in Ψ'_O we let

$$a = b = \alpha , \qquad c = \beta$$

we obtain γ_S. Thus in each term the two factors in the product correspond to conjugate Young diagrams. In the case of a two-particle L-S coupled system this goes over into the simple requirement that a space symmetric function is multiplied by a spin antisymmetric function and vice versa. The latter wave functions belong to the irreducible representations of the permutation group on two objects S_2, and the associated Young diagrams are simply

[1] [2]	[1] [2]
(symmetric)	(antisymmetric)

where each is the conjugate of the other.

CHAPTER SEVEN

MOLECULAR VIBRATIONS AND BRILLOUIN ZONE SYMMETRY

NORMAL MODES OF VIBRATION OF OL_4 TYPE MOLECULE

The application of group theory to the vibrations of symmetrical molecules was done by Wigner (1959) who analyzed the normal modes of tetrahedral methane. We shall illustrate his method by studying the vibrations of an OL_4 type molecule which belongs to the symmetry group C_{4v}.

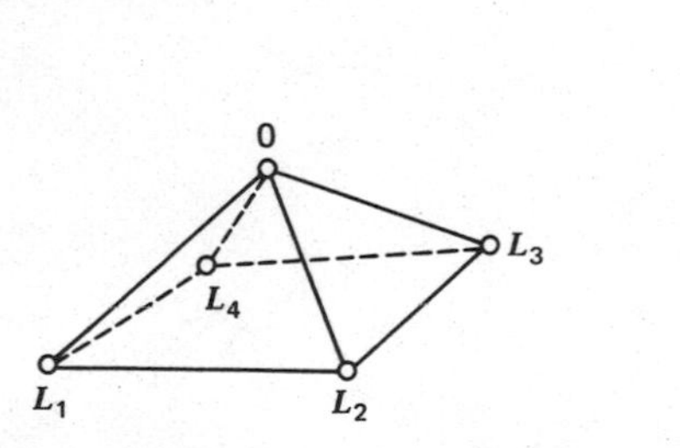

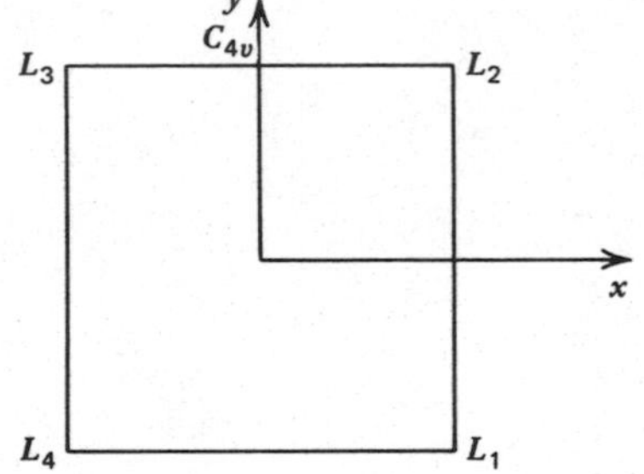

Figure 1. Arrangement of atoms in OL_4 molecule.

The elements of this group of order G = 8 are easily understood with the help of the square $L_1L_2L_3L_4$ in Figure 1. Besides the identity E(≡ E), there are three cyclic rotations C_4(≡ M), C_4^2(≡ N), C_4^3(≡ P) through angles $\psi = 90^\circ$, 180°, and 270°, respectively, about the Z axis passing through the atom O, the reflections σ_x(≡ Q) and σ_y(≡ S) in the vertical planes XZ and YZ, and finally, the two reflections $\sigma_{L_1L_3}$ ($\equiv$ T) and $\sigma_{L_2L_4}$ ($\equiv$ U) in the vertical planes through the diagonals. The Cayley table for this point group is given in Table I.

TABLE I. MULTIPLICATION TABLE OF GROUP C_{4v}

	E	M	N	P	Q	S	T	U
E	E	M	N	P	Q	S	T	U
M	M	N	P	E	U	T	Q	S
N	N	P	E	M	S	Q	U	T
P	P	E	M	N	T	U	S	Q
Q	Q	T	S	U	E	N	M	P
S	S	U	Q	T	N	E	P	M
T	T	S	U	Q	P	M	E	N
U	U	Q	T	S	M	P	N	E

The group has five classes K_1 (E), K_2 (C_4^2), K_3 (C_4, C_4^3), K_4 (σ_x, σ_y), and K_5 ($\sigma_{L_1L_3}$, $\sigma_{L_2L_4}$) and, therefore, five irreducible representations. Two of these are one dimensional, A_1 and A_2 symmetric with respect to the cyclic axis (character = 1); two antisymmetrical one-dimensional representations B_1 and B_2 (character = -1); and a degenerate two-dimensional representation. The subscripts 1 and 2 on A and B indicate whether it is symmetric or antisymmetric with respect to the reflections σ_x and σ_y. Reflections are considered as improper rotations through angle $\phi = 0^{\circ}$.

According to the classical mechanics of elastic vibrations there are 3N - 6 normal modes of vibration of a nonplanar system of N point particles. The molecule OL_4, whose five atoms are arranged as shown in Figure 1, thus has 9 normal modes, and each of these modes belongs to one of the irreducible representations of the symmetry group. To know how many modes have a certain symmetry one has to calculate the compound character $\chi(R)$ of each symmetry operation following the recipe

$$\chi(R) = (\mu_R - 2)(1 + 2\cos\phi) \text{ for proper rotations}$$
$$= \mu_R(-1 + 2\cos\phi) \text{ for improper rotations,} \quad (1)$$

μ_R being the number of atoms left unchanged by the geometric operation O_R. These are shown in Table II.

TABLE II. CHARACTER TABLE

	E	M	N	P	Q	S	T	U
A_1	1	1	1	1	1	1	1	1
A_2	1	1	1	1	-1	-1	-1	-1
B_1	1	-1	1	-1	1	1	-1	-1
B_2	1	-1	1	-1	-1	-1	1	1
E	2	0	-2	0	0	0	0	0
	0°	90°	180°	270°	0°	0°	0°	0°
$2 \cos \phi$	2	0	-2	0	2	2	2	2
$\pm 1 + 2 \cos \phi$	3	1	-1	1	1	1	1	1
μ_R	5	1	1	1	1	1	3	3
$\chi(R)$	9	-1	1	-1	1	1	3	3

Following the usual reduction formula (Equation 37, Chapter Five), relating characters to compound characters, the number of modes of each symmetry type is

$$N_i = \frac{1}{G} \sum_R \chi(R)\ \chi_i(R) \tag{2}$$

where G is the order of the group and $\chi_i(R)$ is the character of the element O_R in the i-th irreducible representation. We thus have

$$\begin{aligned} N_{A_1} &= \frac{1}{8}\left[(9\text{x}1) + (-1\text{x}1) + (1\text{x}1) + (-1\text{x}1) + (1\text{x}1)\right. \\ &\qquad \left. + (1\text{x}1) + (3\text{x}1) + (3\text{x}1)\right] \\ &= 2 \end{aligned} \tag{3}$$

Similar calculations give $N_{A_2} = 0$ (nongenuine vibration), $N_{B_1} = 1$, $N_{B_2} = 2$, and $N_E = 2$. That there are two equal frequencies in each degenerate mode accounts for the nine frequencies of which two are of symmetry A_1, one of B_1, two of B_2, and two of the degenerate E type. Pictures of these modes are to be found in the book by Herzberg (1959). Whether these modes are raman active or infrared active or both is determined by the transformation properties of the dipole moment vector and the polarizability tensor. From the known character table for C_{4v} (Herzberg or Cotton, for instance) we notice that a component of the dipole moment vector (coordinate z) as well as that of the polarizability tensor $(x^2 + y^2)$ transforms as the A_1 representation of the group. The case of the E representation is similar. On the other hand, only components of the polarizability tensor have B_1 and B_2 symmetries. According to Wigner, the A_1 and E type modes are both raman and infrared active whereas B_1 and B_2 type modes are only raman active

SYMMETRY COORDINATES

An orthonormal set of symmetry coordinates (also called symmetry adapted functions) associated with the modes can be found using projection operators that are constructed from the matrix elements of the irreducible representations. Although four of the representations are one-dimensional, the two-dimensional representation matrices are

$$\underset{D(E)}{\begin{pmatrix} 1 & 0 \\ 0 & 1 \end{pmatrix}} \quad \underset{D(M)}{\begin{pmatrix} 0 & -1 \\ 1 & 0 \end{pmatrix}} \quad \underset{D(N)}{\begin{pmatrix} -1 & 0 \\ 0 & -1 \end{pmatrix}} \quad \underset{D(P)}{\begin{pmatrix} 0 & 1 \\ -1 & 0 \end{pmatrix}}$$

$$\underset{D(Q)}{\begin{pmatrix} 1 & 0 \\ 0 & -1 \end{pmatrix}} \quad \underset{D(S)}{\begin{pmatrix} -1 & 0 \\ 0 & 1 \end{pmatrix}} \quad \underset{D(T)}{\begin{pmatrix} 0 & -1 \\ -1 & 0 \end{pmatrix}} \quad \underset{D(U)}{\begin{pmatrix} 0 & 1 \\ 1 & 0 \end{pmatrix}} \qquad (4)$$

With formula 55 of Chapter Three the projection operators are easily made up as

$$P_{A_1} = \frac{1}{8}(E + M + N + P + Q + S + T + U)$$

$$P_{B_1} = \frac{1}{8}(E - M + N - P + Q + S - T - U)$$

$$P_{B_2} = \frac{1}{8}(E - M + N - P - Q - S + T + U)$$

$$P_{E_1} = \frac{1}{4}(E - N + Q - S)$$

$$P_{E_2} = \frac{1}{4}(E - N - Q + S) \quad (5)$$

These operators satisfy well-known relations:

$$P_{A_1} P_{A_1} = P_{A_1}, \quad P_{A_1} P_{B_1} = 0, \text{ etc.}$$

For instance,

$$P_{E_1} P_{E_2} = \frac{1}{16}((E - N)(E - N) - (Q - S)(Q - S) - (E - N)(Q - S) + (Q - S)(E - N))$$

$$= \frac{1}{16}(E - 2N + E - E + N - N - E - Q + S + S - Q + Q + Q - S - S) = 0 \quad (6)$$

As arbitrary functions on which these projection operators operate to generate symmetry coordinates let us choose r_1, r_2, r_3, r_4 and θ_{12}, θ_{23}, θ_{34}, θ_{41}, which are, respectively, the changes in the lengths of the bonds connecting O to each of the L atoms and the changes in the corresponding bond angles as the system vibrates. The effect of the symmetry operations of the group on each of these functions is summarized in Table III. It is easy to establish, using Table III, that

$$P_{A_1}(r_1) = \frac{1}{8}(E + M + N + P + Q + S + T + U)(r_1)$$

$$= \frac{1}{4}(r_1 + r_2 + r_3 + r_4) \quad (7)$$

TABLE III. RESULTS OF GROUP OPERATIONS ON SYMMETRY COORDINATES

	E	M	N	P	Q	S	T	U
r_1	r_1	r_2	r_3	r_4	r_4	r_2	r_3	r_1
r_2	r_2	r_3	r_4	r_1	r_3	r_1	r_2	r_4
r_3	r_3	r_4	r_1	r_2	r_2	r_4	r_1	r_3
r_4	r_4	r_1	r_2	r_3	r_1	r_3	r_4	r_2
θ_{12}	θ_{12}	θ_{23}	θ_{34}	θ_{41}	θ_{34}	θ_{12}	θ_{23}	θ_{41}
θ_{23}	θ_{23}	θ_{34}	θ_{41}	θ_{12}	θ_{23}	θ_{41}	θ_{12}	θ_{34}
θ_{34}	θ_{34}	θ_{41}	θ_{12}	θ_{23}	θ_{12}	θ_{34}	θ_{41}	θ_{23}
θ_{41}	θ_{41}	θ_{12}	θ_{23}	θ_{34}	θ_{41}	θ_{23}	θ_{34}	θ_{12}

and when normalized this becomes $S_{A_1} = \frac{1}{2}(r_1 + r_2 + r_3 + r_4)$.
Likewise, the other symmetry coordinates are

$$S_{B_1} = P_{B_1}(\theta_{12}) \rightarrow \frac{1}{2}(\theta_{12} - \theta_{23} + \theta_{34} - \theta_{41})$$

$$S_{B_2} = P_{B_2}(r_1 + r_2 + r_3) \rightarrow \frac{1}{2}(r_1 - r_2 + r_3 - r_4)$$

$$S_{E_1} = P_{E_1}(\theta_{12} + \theta_{23}) \rightarrow \frac{1}{2}(\theta_{23} - \theta_{41})$$

$$S_{E_2} = P_{E_2}(\theta_{12} + \theta_{23}) \rightarrow \frac{1}{2}(\theta_{12} - \theta_{34}) \tag{8}$$

Here we assume that our basis functions are orthonormal, that is,

$$r_1 r_1 = 1 = \theta_{12} \theta_{12}; \quad r_1 r_2 = 0 = r_1 \theta_{12}$$

and this ensures that the above symmetry coordinates form an orthonormal set. These are by no means unique. It is this set of coordinates that goes into the dynamical analysis involving force constants that ultimately leads to a calculation of the numerical frequencies of these normal modes.

SYMMETRY OF BRILLOUIN ZONES IN SOLIDS

The symmetry properties of wave functions relating to Brillouin zones (B-Z) in crystals have been studied by Bouchaert, Smoluchowski and Wigner (BSW). In this section we discuss the *compatibility relations* between symmetry points in the B-Z which essentially restrict the allowed types of these zones.

The symmetry of a crystal is specified by the space group to which it belongs, a group which includes elements of translations in space as well as rotations and reflections that leave an origin invariant (point group). The Bloch wave function of an electron

$$\Psi_{\vec{k}}(\vec{r}) = e^{i\vec{k}\cdot\vec{r}} U_{\vec{k}}(\vec{r})$$

which generates the representations of the space group is a product of two factors: the phase function $e^{i\vec{k}\cdot\vec{r}}$ that determines what is known as the "star of $\vec{k}$" and the periodic function $U_{\vec{k}}(\vec{r})$ that determines the "small representations" of the "group of the wave vector" $\vec{k}$. The "star of $\vec{k}$" is the figure one obtains when a given wave vector $\vec{k}$ is subjected to all the symmetry operations of the point group. In ascertaining the star of $\vec{k}$ one should note, especially if $\vec{k}$ terminates on a zone boundary, that two points separated by a reciprocal lattice vector $\vec{K}_m$ are considered identical. Those elements of the point group that leave a $\vec{k}$ invariant constitute a subgroup called the "group of the wave vector." Suppose an irreducible representation Γ_j of the point group is decomposed in terms of the irreducible representations of its subgroup. If Γ_i, a given irreducible representation of the latter, occurs in this decomposition, it is said to be compatible with Γ_j. In band theory compatibility relates states that can exist together in a single band.

To illustrate the star of $\vec{k}$ let us study the case of a two-dimensional square zone the length of whose side equals a reciprocal lattice vector (an example to be found in BSW paper and Tinkham's 1975 book). Let our k vector be represented by OE in Figure 2 such that the position OE corresponds to the identity element of the group C_{4v} of the square. The other symmetry operations C_4, C_4^2, C_4^3, σ_x, σ_y, σ_d, $\sigma_{d'}$, take the vector OE into the positions shown in the figure, which does resemble a star. On the other hand, there are no symmetry operations, other than the identity operation, that leave OE alone. Hence the "group of the wave vector" is just this trivial one-element group for the $\vec{k}$ represented by OR.

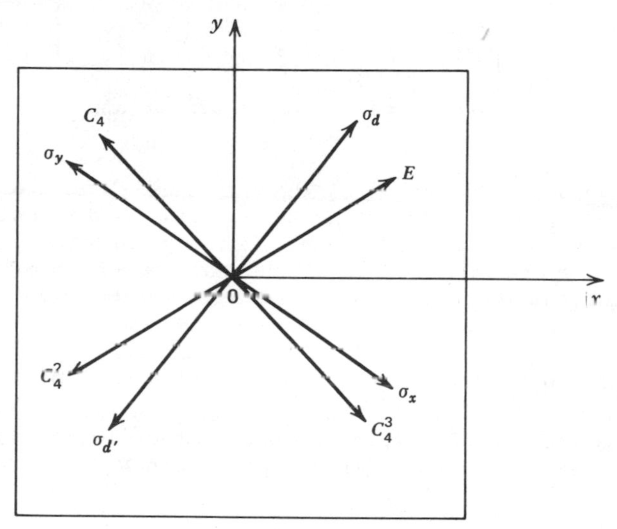

Figure 2

In Figure 3 the k vector is along the X axis and its end point does not touch the zone boundary. The positions to which the vector is carried by the operations C_4 and σ_d, C_4^2 and σ_y, C_4^3 and $\sigma_{d'}$, are pairwise coincident, and the star looks simpler. The point on the axis corresponding to the tip E is akin to what BSW call "Δ". It is easy to see that the symmetry operations of (a) identity and (b) reflection in the XZ plane (σ_x) leave the wave vector invariant while the other operations of the point group alter the orientation of $\vec{k}$. Thus the group of the wave vector is isomorphic to the reflection group C_s. Figures 2 and 3 show two dimensional square zone and results of symmetry operations on $\vec{k}$.

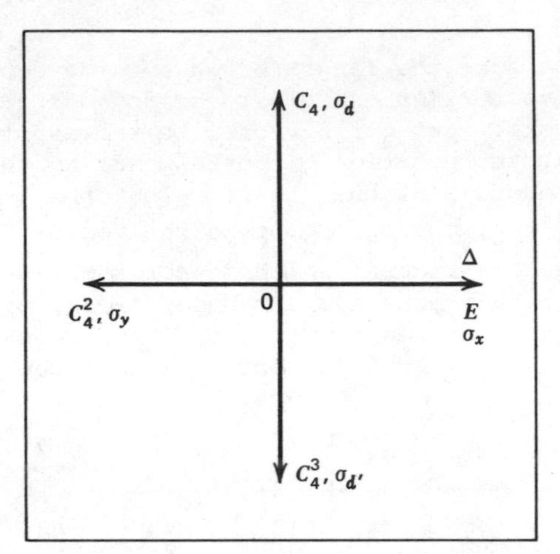

Figure 3

In Figure 4 the star of $\vec{k}$, which is now along a diagonal (still not extending up to the zone boundary) has only four members, as shown. The point E has symmetry corresponding to what is called "Σ" by BSW. It is clear that the group of the wave vector has the two elements, E and σ_d. If the wave vector in Figure 2 extends to the zone boundary as in Figure 5 (the symmetry point E now akin to "Z") the star has only four members because, for instance, the position to which the operation of σ_y carries it is identical to the original position, the two differ- only by a reciprocal lattice vector. Other identical positions are the results of the pairs of operations σ_d, C_4^3; C_4, σ_d; C_4^2,

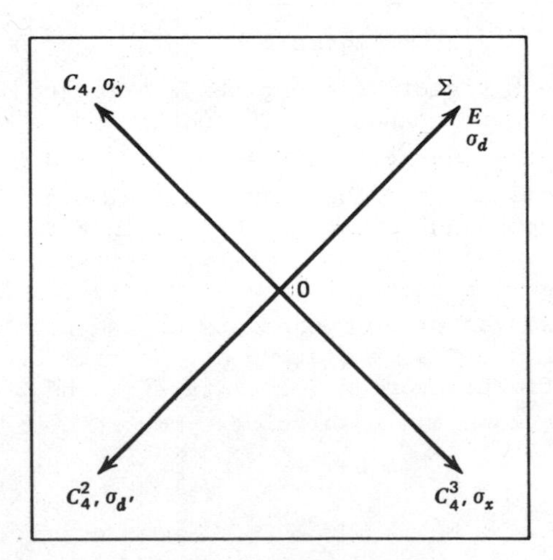

Figure 4

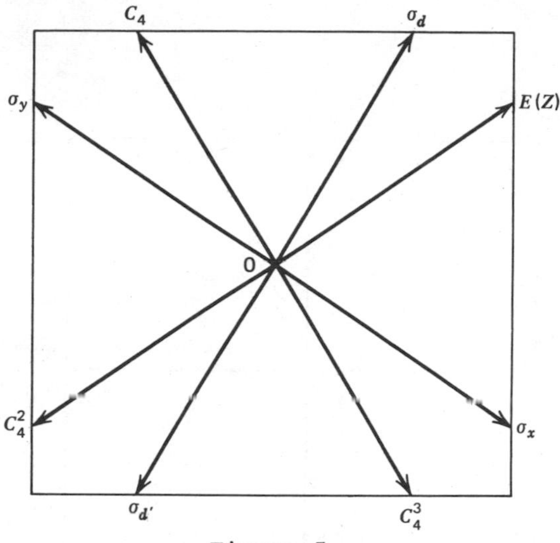

Figure 5

σ_x. The operations E and σ_y leave this invariant and these two constitute the group of the wave vector. Figures 4 and 5 show results of symmetry operations on a different k vector than in earlier figures.

In Figure 6, where the k vector in Figure 3 now extends to the zone boundary (the symmetry corresponding to X of BSW) the star has only two members taking into account the identity of the results of the sets of operations E, C_4^2, σ_y, σ_x; C_4, C_4^3, σ_d, $\sigma_{d'}$. The symmetry operations E, σ_x, C_4^2, σ_y leave the wave vector invariant, and these are the elements of the group of the wave vector. The wave vector along a diagonal in Figure 4 extends to the zone boundary in Figure 7 (symmetry of its tip now called M), and all the orientations into which the symmetry operations carry it are now identical. The star thus degenerates into the original wave vector. On the other hand, all the symmetry operations of the group leave it invariant (modulo $\vec{K}_m$), and hence the group of the wave vector is the full symmetry group C_{4v}. The symmetry at the point C (called Γ) corresponds to the group of the wave vector $\vec{k} = 0$. It is to be noted that the symmetry points Γ, Δ, Σ, M, X, Z of BSW are designated with respect to the full cubic group of a Brillouin zone, whereas we have here adapted these to a two-dimensional section of the cube. Figures 6 and 7 show results of symmetry operation on a K vector that touches the zone boundaries.

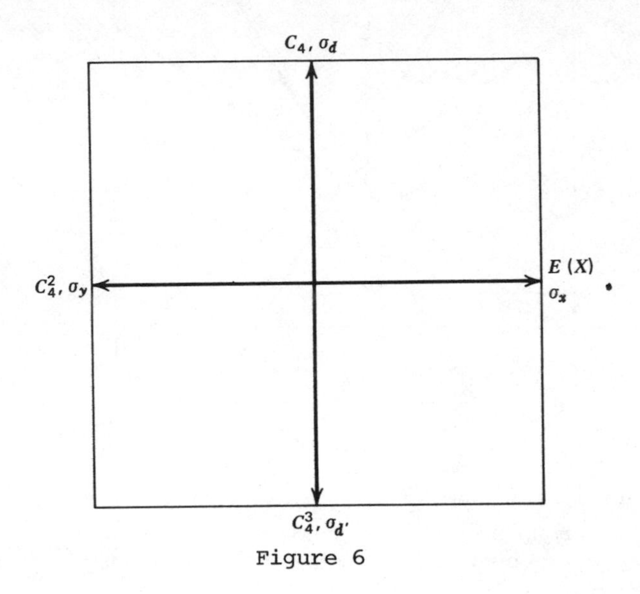

Figure 6

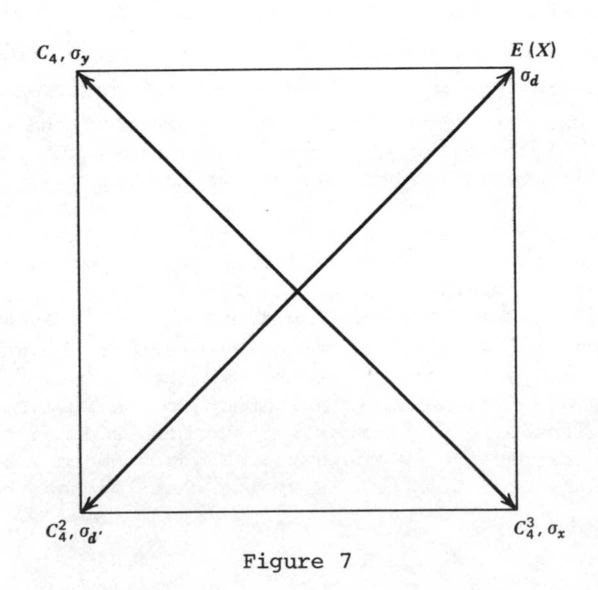

Figure 7

To understand the compatibility relations we consider a B-Z with the symmetry of the simple cubic lattice as in Figure 8. The wave vector $\vec{k} = 0$ (Γ) has the full symmetry O_h, as does the wave vector Γ R in the figure. To establish these compatibility relations between points Γ and Λ consider a k vector which lies at an

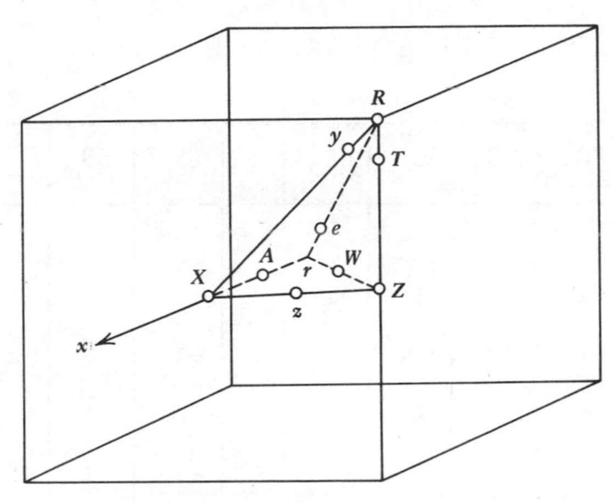

Figure 8

intermediate position along the body diagonal (line joining Γ and Λ). This vector is invariant to three rotations C_3, C_3^2, C_3^3 (identity) about this line as the cyclic axis, and reflections in the diagonal planes of the cube passing through this line. These are the six symmetry operations that are elements of O_h and that also leave the wave vector Γ Λ invariant. It is easy to see that these six operations are elements of the group C_{3v}. The charácters of these elements in the irreducible representations of C_{3v} and the characters of the elements of O_{3h} in some of its irreducible representations are given in Table IV.
With the use of the above characters in Table IV of the reducible representations in the well-known reduction formula (Equation 37, Chapter Five) a trivial calculation shows that

$$A_{1g} = A_1 \qquad E_g = E \qquad T_{1g} = A_2 + E$$

$$A_{1u} = A_2 \qquad E_u = E \text{ and } T_{1u} = A_1 + E\ .$$

We thus conclude that A_{1g} is compatible with A_1, E_g with E, A_{1u}

TABLE IV. CHARACTERS OF THE SYMMETRY ELEMENTS OF O_{3h}

	E	$8C_3$	$6C_4$	$3C_4^2$	$6C_2$	i	$8S_6$	$6S_4$	$3\sigma_v$	$6\sigma_d$
O_h										
A_{1g}	1	1	1	1	1	1	1	1	1	1
E_g	2	-1	0	2	0	2	-1	0	2	0
T_{1g}	3	0	1	-1	-1	3	0	1	-1	-1
A_{1u}	1	1	1	1	1	-1	-1	-1	-1	-1
E_u	2	-1	0	2	0	-2	1	0	-2	0
T_{1u}	3	0	1	-1	-1	-3	0	-1	1	1
C_{3v}										
A_1	1	1								1
A_2	1	1								-1
E	2	-1								0

with A_2, E_u with E, and further, that T_{1g} is compatible with A_2 and E and T_{1u} with A_1 and E. There is then some sort of continuity between the symmetry orbitals at Λ and at Γ. A comprehensive discussion of these relations is to be found in the original work of BSW and Slater (1972).

CHAPTER EIGHT

APPLICATIONS IN PARTICLE PHYSICS

UNITARY IRREDUCIBLE REPRESENTATIONS OF SU(2)

Consider the transformation

$$U(a,b): \begin{pmatrix} u' \\ v' \end{pmatrix} = U(a,b) \begin{pmatrix} u \\ v \end{pmatrix} = \begin{pmatrix} a^* & -b \\ b^* & a \end{pmatrix} \begin{pmatrix} u \\ v \end{pmatrix} \tag{1}$$

with $|a|^2 + |b|^2 = 1$. U is thus a unitary matrix, $U^{-1} = U^+$, and is an element of the SU(2) group. To obtain the unitary irreducible representations consider the basis polynomials (functions of u and v called monomials, also called tensors)

$$f_J^m = \frac{u^{J+m} v^{J-m}}{\sqrt{(J+m)!\,(J-m)!}}, \quad m = -J, \cdots, +J$$

and

$$U(a,b) f_J^m = f'^m_J = f_J^m \left(U^{-1} \begin{pmatrix} u \\ v \end{pmatrix} \right) = \frac{(au+bv)^{J+m} \, (-b^*u+a^*v)^{J-m}}{\sqrt{(J+m)!\,(J-m)!}} \tag{2}$$

As shown in Hamermesh's text (1962) a straightforward binomial expansion leads to the relation

$$U(a,b)f_J^m = \sum_{m'} f_J^{m'} D_{m'm}^J(a,b)$$

where the representation matrix $D_{m'm}^J$ is explicitly

$$\sum_k \frac{\sqrt{(J+m)!\ (J-m)!\ (J+m')!\ (J-m')!}}{(J+m-k)!\ k!(J-m'-k)!\ (m'-m+k)!} a^{J+m-k} a^{*J-m'-k} b^k (-b^*)^{m'-m+k} \tag{3}$$

We now show that the representation is unitary. The normalization of the f_J^m was chosen so that

$$\sum_{m=-J}^{J} f_J^m (f_J^m)^* = \sum_{m=-J}^{J} \frac{1}{(J+m)!\ (J-m)!} |u|^{2(J+m)} |v|^{2(J-m)}$$

$$= \sum_{k=0}^{2J} \frac{|u|^{2k}|v|^{2(2J-k)}}{k!(2J-k)!} = \frac{(|u|^2 + |v|^2)^{2J}}{(2J)!}, \tag{4}$$

by binomial expansion. Similarly,

$$\sum_{m=-J}^{J} f_J^m (f_J'^m)^* = \sum_{m=-J}^{J} |U(a,b)f_J^m|^2$$

$$= \sum_{m=-J}^{J} \frac{|au + bv|^{2(J+m)} |-b^*u + a^*v|^{2(J-m)}}{(J+m)!\ (J-m)!}$$

$$= \frac{(|au+bv|^2 + |-b^*u+a^*v|^2)^{2J}}{(2J)!}$$

$$= \frac{(|u|^2 + |v|^2)(|a|^2 + |b|^2)}{(2J)!}$$

$$= \frac{(|u|^2 + |v|^2}{(2J)!} = \sum_{m=-J}^{J} |f_J^m|^2 \tag{5}$$

Therefore,

$$\sum_{m=-J}^{J} \sum_{m'} f_J^{m'} D^J_{m'm} \sum_{m''} (f_J^{m''})^* (D^J_{m''m})^* = \sum_{m=-J}^{J} f_J^m (f_J^m)^* \tag{6}$$

Since the $(2J + 1)^2$ functions $f_{m'}(f_{m''})^*$ are linearly independent, the representation matrices must satisfy

$$\sum_{m=-J}^{J} D^J_{m'm} (D^J_{m''m})^* = \delta_{m'm''} \tag{7}$$

That is, $DD^+ = I$ and the representations are unitary. So we have a unitary representation. Is it irreducible?

<u>Set b = 0.</u> Try for diagonal D.

$k = 0$, $m = m'$ for non-zero term

$$D^J_{m'm}(a,0) = \frac{\delta_{mm'}(J+m)!\ (J-m)!}{(J+m)!\ (J-m)!} a^{J+m} (a^*)^{J-m}$$

and

$$|a| = 1,\ a = e^{i\alpha/2}$$

$$D^J_{m'm}(a,0) = \delta_{mm'} e^{i\alpha/2(J+m)} e^{-i\alpha/2(J-m)}$$

and also

$$= \delta_{mm'} e^{i\alpha m} \tag{8}$$

For example,

$$D^{\frac{1}{2}}(a,0) = \begin{pmatrix} e^{-i\alpha/2} & 0 \\ 0 & e^{+i\alpha/2} \end{pmatrix}$$

Now any matrix A that commutes with all $D^J(a,0)$ is diagonal; that is,

$$A_{mm'} = \delta_{mm'} A_{mm}$$

Consider $D^J_{Jm}(a,b)$. Only k = 0 in sum gives a nonzero contribution:

$$D^J_{Jm}(a,b) = \sqrt{\frac{(2J)!}{(J+m)!\,(J-m)!}}\; a^{J+m}(-b^*)^{J-m} \qquad (9)$$

So, in general,

$$D^J_{Jm}(a,b) \neq 0$$

Now, if A commutes with all $D^J(a,b)$, then

$$(AD^J)_{Jm} = (D^J A)_{Jm}$$

$$\therefore \quad A_{JJ} D^J_{Jm} = D^J_{Jm} A_{mm} \qquad \text{for all } m$$

and

$$A_{mm} = A_{JJ} \qquad \text{for all } m$$

$$\therefore \quad A = A_{JJ} I \qquad (10)$$

that is, a matrix A which commutes with all matrices of the representation $D^J(a,b)$ must be a multiple of the unit matrix. Hence, by Schur's lemma, the representations $D^J(a,b)$ are irreducible.

CLEBSCH-GORDAN COEFFICIENTS FOR SU(2)

In an earlier section we discussed the irreducible representations of an SU(2) group that has three infinitesimal generators J_1, J_2, J_3 and the Casimir invariant $J^2 = J_1^2 + J_2^2 + J_3^2$. The $(2j + 1)$ dimensional representations $D^{(j)}$ are generated by the basis functions $|jm\rangle$ where the labels are, respectively, the eigenvalues of J^2 $(j(j+1))$ and $J_3(m)$. The electron spin in quantum mechanics is an outstanding example of SU(2) symmetry. Consider two different irreducible representations $D^{(j_1)}$ and $D^{(j_2)}$

(say, for instance, two electron spins). $D^{(j_1)} \otimes D^{(j_2)}$, a direct product of the two representations, is, in general, reducible. The Clebsch-Gordan (C-G) theorem essentially relates to the reduction of this representation:

$$\text{(a)} \quad D^{(j_1)} \otimes D^{(j_2)} = \sum_J D^{(J)} \quad \text{where } |j_1-j_2| \le J \le (j_1+j_2)$$

$$\text{(b)} \quad |J\ M> = \sum_{m_1 m_2} <j_1 m_1\ j_2 m_2|J\ M> |j_1 m_1> |j_2 m_2> \qquad (11)$$

where $|j_1 m_1>$ are the basis functions of the irreducible representations $D^{(j_1)}$ and similarly for $|j_2 m_2>$ $\cdot$ $|J\ M>$ are the basis functions of the irreducible representations $D^{(J)}$. $<j_1 j_2 m_1 m_2|JM>$ are real numbers called C-G (or Wigner) coefficients.

This corresponds to the quantum mechanical vector addition of two commuting angular momentum vectors $\vec{J}_1 + \vec{J}_2 = \vec{J}$. The derivation of these coefficients is given in several textbooks and Wigner's (1959) well-known classic. Here we give instead a brief demonstration by induction for the case $j_2 = \frac{1}{2}$.

According to the C-G theorem we have

$$|J = j_1+\tfrac{1}{2}, M=m> = \sqrt{\frac{j_1+m+\frac{1}{2}}{2j_1+1}}\ |j_1, m_1 = m-\tfrac{1}{2}> |j_2 = \tfrac{1}{2}, m_2 = \tfrac{1}{2}>$$

$$+ \sqrt{\frac{j_1-m+\frac{1}{2}}{2j_1+1}}\ |j_1, m_1 = m+\tfrac{1}{2}> |j_2 = \tfrac{1}{2}, m_2 = -\tfrac{1}{2}>$$

$$|j_1 - \tfrac{1}{2}, m> = -\sqrt{\frac{j_1-m+\frac{1}{2}}{2j_1+1}}\ |j_1, m-\tfrac{1}{2}> |\tfrac{1}{2}\ \tfrac{1}{2}>$$

$$+ \sqrt{\frac{j_1+m+\frac{1}{2}}{2j_1+1}}\ |j_1, m+\tfrac{1}{2}> |\tfrac{1}{2}\ -\tfrac{1}{2}> \qquad (12)$$

It is true for $m = j_1+\frac{1}{2}$: $|j_1+\frac{1}{2}, j_1+\frac{1}{2}\rangle = |j_1 j_1\rangle\, |\frac{1}{2}\,\frac{1}{2}\rangle$. Assume it is true for m. We now show it is true for m - 1. Once a basis function of a representation $D^{(j)}$ is known, its partners can be generated by means of the step-up and step-down operators J_+ and J_- that satisfy

$$J_\pm\, |j_1 j_3\rangle = \sqrt{(j \mp j_3)(j \pm j_3 + 1)}\; |j_1 j_3 \pm 1\rangle$$

Similar relations are well-known in quantum mechanics where, for instance, $|1\ m\rangle$ will be a spherical harmonic when J refers to the orbital angular momentum. Incidentally, it is important to note that SU(2) is the covering group for the three-dimensional rotation group O(3) to which the spherical harmonics belong. Operating with $(J_1+J_2)_- = J_-$ on the state $|j_1 + \frac{1}{2}, m\rangle$ we have

$$J_-|j_1+\tfrac{1}{2},m\rangle = \sqrt{\frac{j_1+m+\frac{1}{2}}{2j_1+1}}\left\{\sqrt{(j_1+m-\tfrac{1}{2})(j_1-m+3/2)}\;|j_1,m-3/2\rangle|\tfrac{1}{2},\tfrac{1}{2}\rangle\right.$$

$$\left. + |j_1,m-\tfrac{1}{2}\rangle|\tfrac{1}{2},-\tfrac{1}{2}\rangle\right\} + \left\{\sqrt{\frac{j_1-m+\frac{1}{2}}{2j_1+1}}\sqrt{(j_1+m+\tfrac{1}{2})(j_1-m+\tfrac{1}{2})}\right.$$

$$\left. \times\, |j_1,m-\tfrac{1}{2}\rangle\;|\tfrac{1}{2},-\tfrac{1}{2}\rangle\right\} = \sqrt{\frac{j_1+m+\frac{1}{2}}{2j_1+1}}\left\{(j_1+m-\tfrac{1}{2})\right.$$

$$\times\,(j_1-m+3/2)\,|j_1,m-3/2\rangle|\tfrac{1}{2},\tfrac{1}{2}\rangle + (j_1-m+3/2)$$

$$\left. \times\, |j_1,m-\tfrac{1}{2}\rangle|\tfrac{1}{2},-\tfrac{1}{2}\rangle\right\} \qquad (13)$$

The representations being unitary, the basis functions must be ortho-normal. We normalize the above by multiplying it with the factor

$$\frac{1}{\sqrt{(j_1+m+\frac{1}{2})(j_1-m+3/2)}}$$

and obtain

$$|j_1+\tfrac{1}{2},m-1\rangle = \sqrt{\frac{j_1+m-\frac{1}{2}}{2j_1+1}}\ |j_1, m-3/2\rangle\ |\tfrac{1}{2}\ \tfrac{1}{2}\rangle$$

$$+ \sqrt{\frac{j_1-m+3/2}{2j_1+1}}\ |j_1,m-\tfrac{1}{2}\rangle,\ |\tfrac{1}{2},-\tfrac{1}{2}\rangle \qquad (14)$$

and this means that the result is true for m - 1, and therefore true for all m. In particular when $m = j_1 - \frac{1}{2}$ we have

$$|j_1+\tfrac{1}{2},\ j_1-\tfrac{1}{2}\rangle = \sqrt{\frac{2j_1}{2j_1+1}}\ |j_1,\ j_1-1\rangle\ |\tfrac{1}{2},\tfrac{1}{2}\rangle$$

$$+ \sqrt{\frac{1}{2j_1+1}}\ |j_1,j_1\rangle\ |\tfrac{1}{2},-\tfrac{1}{2}\rangle \qquad (15)$$

Now for the case $j = j_1 - \frac{1}{2}$ we note that the basis function $|j_1-\frac{1}{2},j_1-\frac{1}{2}\rangle$ must be normalized and orthogonal to the above state. Equation (12) gives with $m = j_1 - \frac{1}{2}$

$$|j_1-\tfrac{1}{2},j_1-\tfrac{1}{2}\rangle = -\sqrt{\frac{1}{2j_1+1}}\ |j_1,j_1-1\rangle\ |\tfrac{1}{2},\tfrac{1}{2}\rangle$$

$$+ \sqrt{\frac{2j_1}{2j_1+1}}\ |j_1,j_1\rangle\ |\tfrac{1}{2},-\tfrac{1}{2}\rangle \qquad (16)$$

which is indeed orthogonal to $|j_1+\frac{1}{2},\ j_1-\frac{1}{2}\rangle$ and is normalized. Hence the equation is correct for $m = j_1 - \frac{1}{2}$. By going through a process of induction similar to the one for $j = j_1 + \frac{1}{2}$ the proof is readily established.

ISOSPIN - CLEBSCH-GORDAN COEFFICIENTS

Consider the reactions:

(1) $\pi^+ + p \rightarrow \pi^+ + p$

(2) $\pi^- + p \rightarrow \pi^- + p$

(3) $\pi^- + p \rightarrow \pi^o + n$

The C-G expansion is

$$|JM\rangle = \sum_{m_1+m_2=M} \langle 1\ m_1\ \tfrac{1}{2}\ m_2|JM\rangle |1\ m_2\rangle|\tfrac{1}{2}\ m_2\rangle$$

$$\left[|JM\rangle = \sum_{m_1+m_2=M} C^{JM}_{1m_1\frac{1}{2}m_2} |1,m_1\rangle|\tfrac{1}{2},\ m_2\rangle \right]$$

$$|3/2\ 3/2\rangle = |1,\ 1\rangle|\tfrac{1}{2},\ \tfrac{1}{2}\rangle$$

$$|3/2\ 1/2\rangle = \sqrt{1/3}\ |\bar{1},1\rangle\ |\tfrac{1}{2}\ -\tfrac{1}{2}\rangle + \sqrt{2/3}\ |1,0\rangle\ |\tfrac{1}{2}\tfrac{1}{2}\rangle$$

$$|3/2\ -\ 1/2\rangle = \sqrt{1/3}\ |1,\ -1\rangle\,|1/2,\ 1/2\rangle + \sqrt{2/3}\ |1,0\rangle\,|1/2\ \ -1/2\rangle$$

$$|3/2\ -\ 3/2\rangle = |1\ -\ 1\rangle\ |1/2\ -\ 1/2\rangle$$

$$-\ |1/2,\ 1/2\rangle = \sqrt{1/3}\ |1/2,\ 1/2\rangle\ |1,\ 0\rangle - \sqrt{2/3}\ |1/2\ -\ 1/2\rangle\ |1,\ 1\rangle$$

$$|1/2,\ -1/2\rangle = \sqrt{1/3}\ |1/2\ -\ 1/2\rangle\ |10\rangle - \sqrt{2/3}\ |1/2\ 1/2\rangle\ |1,\ -1\rangle \qquad (17)$$

Identifying,

$$|1\ 1\rangle = |\pi^+\rangle$$

$$|1\ 0\rangle = |\pi^o\rangle$$

$$|1\ -1\rangle = |\pi^-\rangle$$

$$|1/2\ 1/2\rangle = |p\rangle$$

$$|1/2\ -1/2\rangle = |n\rangle \tag{18}$$

we obtain

$$|3/2,\ 3/2\rangle = |p,\ \pi^+\rangle$$

$$|3/2,\ 1/2\rangle = \sqrt{1/3}|n,\ \pi^+\rangle + \sqrt{2/3}|p,\ \pi^0\rangle$$

$$|3/2,\ -1/2\rangle = \sqrt{1/3}|\ p,\ \pi^-\rangle + \sqrt{2/3}|\ n,\ \pi^0\rangle \tag{19}$$

$$|3/2,\ -3/2\rangle = |n,\ \pi^-\rangle$$

$$|1/2,\ 1/2\rangle = \sqrt{2/3}\ |n,\ \pi^+\rangle - \sqrt{1/3}\ |p,\ \pi^0\rangle$$

$$|1/2,\ -1/2\rangle = \sqrt{1/3}\ |n,\ \pi^0\rangle - \sqrt{2/3}\ |p,\ \pi^-\rangle$$

$$|p,\ \pi^+\rangle = |3/2,\ 3/2\rangle$$

$$|p,\ \pi^-\rangle = \sqrt{1/3}\ |\ 3/2,\ -1/2\rangle - \sqrt{2/3}\ |\ 1/2,\ -1/2\rangle$$

$$|n,\ \pi^0\rangle = \sqrt{2/3}\ |\ 3/2,\ -1/2\rangle + \sqrt{1/3}\ |\ 1/2,\ -1/2\rangle \tag{20}$$

By SU(2) invariance of (isospin conservation in) strong interactions

$$\frac{\langle\pi^+p|T|\pi^+p\rangle}{\langle\pi^-p|T|\pi^-p\rangle} = \frac{T^{(3/2)}}{1/3T^{(3/2)} + 2/3T^{(1/2)}}$$

and

$$\frac{\langle\pi^+p|T|\pi^+p\rangle}{\langle\pi^0n|T|\pi^-p\rangle} = \frac{T^{(3/2)}}{\sqrt{2/3}\ T^{(3/2)} - \sqrt{2/3}\ T^{(1/2)}}$$

At low energy we may take

$$T^{(1/2)} = 0,$$

and

$$\sigma(\pi^+ p \to \pi^+ p) : \sigma(\pi^- p \to \pi^- p) : \sigma(\pi^- p \to \pi^0 n)$$

$$= |\langle\pi^+ p|T|\pi^+ p\rangle|^2 : |\langle\pi^- p|T|\pi^- p\rangle|^2 : |\langle\pi^0 n|T|\pi p\rangle|^2$$

$$= 1 : (1/3)^2 . \left(\frac{\sqrt{2}}{3}\right)^2$$

$$= 9:1:2 \qquad (21)$$

In general, for $T^{(1/2)} \neq 0$

$$\sqrt{2} \langle\pi^0 n|T|\pi^- p\rangle + \langle\pi^- p|T|\pi^- p\rangle = \langle\pi^+ p|T|\pi^+ p\rangle$$

SU(3) AND PARTICLE PHYSICS

SU(3) is the group of transformations $\psi'_a = U_{ab}\psi_b$ where U is any unitary, unimodular 3x3 matrix with determinant $||U|| \neq 0$. As discussed earlier this is the group of the three-dimensional isotropic harmonic oscillator in quantum mechanics of which Schiff gives an account in his text. In terms of Lie's infinitesimal generators U is given by

$$U = \exp\{i \sum_{i=1}^{8} \varepsilon_k F_k\}$$

where $F_k \equiv \frac{1}{2}\lambda_k$ are the infinitesimal generators. An explicit matrix form for the λ_k's, due to Gell-Mann, in which λ_3 and λ_8 are diagonal is

$$\lambda_1 = \begin{pmatrix} 0 & 1 & 0 \\ 1 & 0 & 0 \\ 0 & 0 & 0 \end{pmatrix} \qquad \lambda_2 = \begin{pmatrix} 0 & -i & 0 \\ i & 0 & 0 \\ 0 & 0 & 0 \end{pmatrix} \qquad \lambda_3 = \begin{pmatrix} 1 & 0 & 0 \\ 0 & -1 & 0 \\ 0 & 0 & 0 \end{pmatrix}$$

$$\lambda_4 = \begin{pmatrix} 0 & 0 & 1 \\ 0 & 0 & 0 \\ 1 & 0 & 0 \end{pmatrix} \qquad \lambda_5 = \begin{pmatrix} 0 & 0 & -i \\ 0 & 0 & 0 \\ i & 0 & 0 \end{pmatrix} \qquad \lambda_6 = \begin{pmatrix} 0 & 0 & 0 \\ 0 & 0 & 1 \\ 0 & 1 & 0 \end{pmatrix}$$

$$\lambda_7 = \begin{pmatrix} 0 & 0 & 0 \\ 0 & 0 & -i \\ 0 & i & 0 \end{pmatrix} \qquad \lambda_8 = \begin{pmatrix} \frac{1}{\sqrt{3}} & 0 & 0 \\ 0 & \frac{1}{\sqrt{3}} & 0 \\ 0 & 0 & -\frac{2}{\sqrt{3}} \end{pmatrix} \tag{22}$$

The usual structure relations satisfied by these generators are

$$[F_\alpha, F_\beta] = \sum_\gamma C^\gamma_{\alpha\beta} F_\gamma = i \sum_\gamma f_{\alpha\beta\gamma} F_\gamma .$$

It is more convenient to express the structure constants in terms of $f_{\alpha\beta\gamma}$ rather than $C^\gamma_{\alpha\beta}$. The nonvanishing $f_{\alpha\beta\gamma}$ are

$$\begin{array}{lll} f_{123} = 1 & f_{257} = \frac{1}{2} & f_{678} = \frac{\sqrt{3}}{2} \\ f_{147} = \frac{1}{2} & f_{345} = \frac{1}{2} & \\ f_{156} = -\frac{1}{2} & f_{367} = -\frac{1}{2} & \\ f_{246} = \frac{1}{2} & f_{458} = \frac{\sqrt{3}}{2} & \end{array} \tag{23}$$

The $f_{\alpha\beta\gamma}$ are odd under permutation of any two indices. We define the following standard combinations of generators. We now use the usual notation $T_\pm$, T_3 for isospin instead of $J_\pm$, J_3

$$T_\pm = F_1 \pm iF_2 \qquad U_\pm = F_6 \pm iF_7 \qquad V_\pm = F_4 \pm iF_5$$

$$T_3 = F_3 \qquad Y = \frac{2}{\sqrt{3}} F_8 \tag{24}$$

The commutation relations satisfied by these operators can easily be derived. For example,

$$[T_3, T_\pm] = [F_3, F_1 \pm iF_2] = [F_3, F_1] \pm i [F_3, F_2] = iF_2$$

$$\pm F_1 = \pm T_\pm$$

Similarly,

$$[Y, T_\pm] = 0 = [T_3, Y], [T_3, U_\pm] = \mp \frac{1}{2} U_\pm$$

$$[Y, U_\pm] = \pm U_\pm$$

$$[U_+, U_-] = \frac{3}{2} Y - T_3 = 2U_3$$

$$[V_+, V_-] = \frac{3}{2} Y + T_3 = 2V_3, \text{ etc.} \tag{25}$$

The generators have been chosen so that T_3 and Y are diagonal with eigenvalues t_3 and y. The states Ψ are labeled with t_3 and y. The action of the shift operators $T_\pm$, $U_\pm$, and $V_\pm$ on the state $\Psi(t_3', y')$ is illustrated on the two-dimensional grid in Figure 1. The action of $T_\pm$ is determined by the commutation relations

$$[T_3, T_\pm] = \pm T_\pm \quad \text{and} \quad [Y, T_\pm] = 0,$$

that of $U_\pm$ by

$$[T_3, U_\pm] = \mp \frac{1}{2} U_\pm \quad \text{and} \quad [Y, U_\pm] = \pm U_\pm$$

and that of $V_\pm$ by

$$[T_3, V_\pm] = \pm \frac{1}{2} V_\pm \quad \text{and} \quad [Y, V_\pm] = \pm V_\pm \tag{26}$$

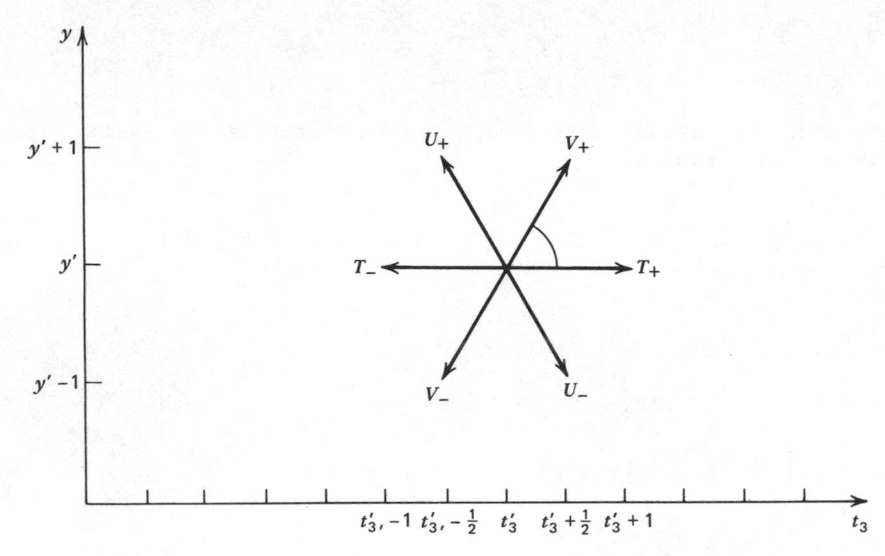

Figure 1. Results of shift operations on the state $\Psi(t_3',y')$.

One can generate all the states of an irreducible representation by repeated application of the shift operators to any one of them. The irreducible representations can be labeled by the pair of integers (p,q). If one begins at the unique state of maximum t_3, Ψ_{max} the boundary of distribution of occupied sites (states) is p steps long in the -120° direction (i.e., $V_-^{p+1}\Psi_{max} = 0$) and $V_-^p \Psi_{max}$ is proportional to state at next corner, moving clockwise along the boundary from Ψ_{max}. If one now heads in the $-t_3$ direction, there are q steps until the next corner in the boundary is reached (i.e., $T_-^{q+1}(V_-^p \Psi_{max}) = 0$) and $T_-^q(V_-^p \Psi_{max})$ is proportional to the state at the second corner, clockwise along the boundary from Ψ_{max} (the boundary is always convex).

The multiplicity (dimensionality) of the representation (p,q) is $N = \frac{1}{2}(p+1)(q+1)(p+q+2)$. This can be seen as follows. Consider the case $p \geq q$. The boundary is, in general, a six-sided figure with sides of length p,q,p,q,p,q as one goes around clockwise from Ψ_{max} and each with interior angles of 120°. There is one state at each site on the boundary. As one moves in from the boundary, each successive six-sided figure has the length of each side reduced by one, and the number of states at each site increased by one. This continues until the short sides are reduced to zero and one has an equilateral triangle with sides of length (p - q) and multiplicity (q + 1). As one continues inward now, the multiplicity remains (q + 1).

The number of states on the boundary and inside the triangle is

$$(q+1) \sum_{k=1}^{p-q+1} k = \tfrac{1}{2}(q+1)(p-q+1)(p-q+2) \tag{27}$$

The number of states on the six-sided figures is given by

$$3 \sum_{k=0}^{q-1} (q-k)(p-q+2k+2)$$

Therefore,

$$N = \tfrac{1}{2}(q+1)(p-q+1)(p-q+2) + 3 \sum_{k=0}^{q-1} (q-k)(p-q+2k+2)$$

$$= \tfrac{1}{2}(p+1)(q+1)(p-q+2) - \tfrac{1}{2}q(q+1)(p-q+2) + 3(p-q+2)\tfrac{1}{2}(q+1)$$

$$+ 6 \sum_{k=0}^{q-1} (q-k)k \tag{28}$$

and since

$$\sum_{k=1}^{q-1} (q-k)k = (q+1)q(q-1)/6$$

we readily get

$$N = \tfrac{1}{2}(p+1)(q+1)(p+q+2).$$

This result, symmetric under interchange of p and q is also valid for $p < q$.

If one now applies U_+ to the state of maximum t_3, Ψ_{max}, one moves along the boundary in the counterclockwise direction reaching the next corner, Ψ' after q steps. These q + 1 states form a U-spin multiplet of which $\Psi_{max} = |U = q/2, u_3 = - q/2\rangle$ is the lowest state and $\Psi' = |U = q/2, u_3 = + q/2\rangle$ is the highest state. Ψ' carries the maximum value of y that occurs in the representation y_{max}, since continuing counterclockwise from Ψ'

the (convex) boundary runs parallel to the t_3 axis and then turns downward to smaller values of y, but u_3, y, and t_3 are not independent: $3/2\, y - t_3 = 2u_3$. Applying this equation to $\Psi'(t_3 = p/2,\ u_3 = q/2)$ we find

$$Y_{max} = \frac{4}{3}\left(\frac{q}{3}\right) + \frac{2}{3}\left(\frac{p}{2}\right) = \frac{1}{3}(p + 2q)$$

(The eigenvalue of Y for Ψ_{max} is $(1/3)(p - q)$). One can also find the value of t_3 carried by Ψ_{max}, $(t_3)_{max}$ and hence the largest value of t, t_{max}, which occurs in the representation. $t_{max} = (t_3)_{max} = \frac{1}{2}(p+q)$, since in moving back from Ψ' to Ψ_{max} the value of t_3 increases by $\frac{1}{2}$ for each of the q steps.

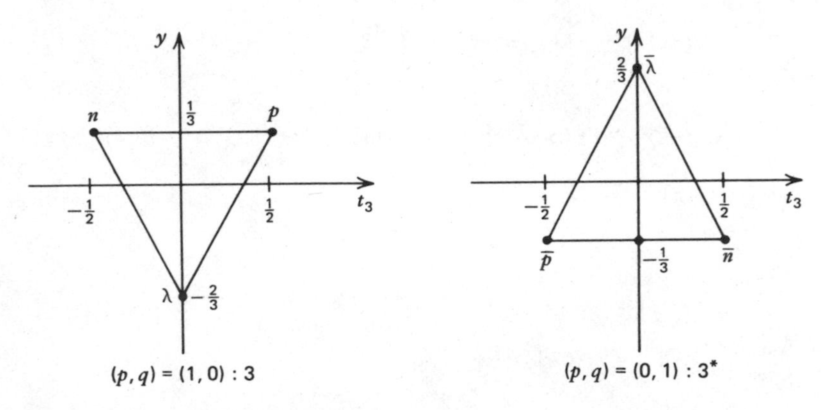

Figure 2. The fundamental triplet representations.

One makes here the usual association of the three (3) representation with the quarks and the three star (3*) representation with the antiquarks, y with hypercharge y = B + S (B is the baryon number and S the strangeness), and t_3 with the third component of isospin. The formula $y_{max} = (p + 2q)/3$ gives $y_{max} = 1/3$ for the three representation, $y_{max} = 2/3$ for the three star representation. Using $t_{max} = \frac{1}{2}(p+q)$ one obtains $t_{max} = \frac{1}{2}$ for both the three and three star. Using the Gell-Mann-Nishijima relation for the charge (in units of the proton charge) $Q = t_3 + \frac{1}{2}y$ we obtain the usual quantum numbers for the quarks and antiquarks:

	B	t	t_3	y	S	Q
p	1/3	1/2	1/2	1/3	0	2/3
n	1/3	1/2	-1/2	1/3	0	-1/3
λ	1/3	0	0	-2/3	-1	-1/3
$\bar{p}$	-1/3	1/2	-1/2	-1/3	0	-2/3
$\bar{n}$	-1/3	1/2	1/2	-1/3	0	1/3
$\bar{\lambda}$	-1/3	0	0	2/3	1	1/3

These representations are shown in Figure 2.

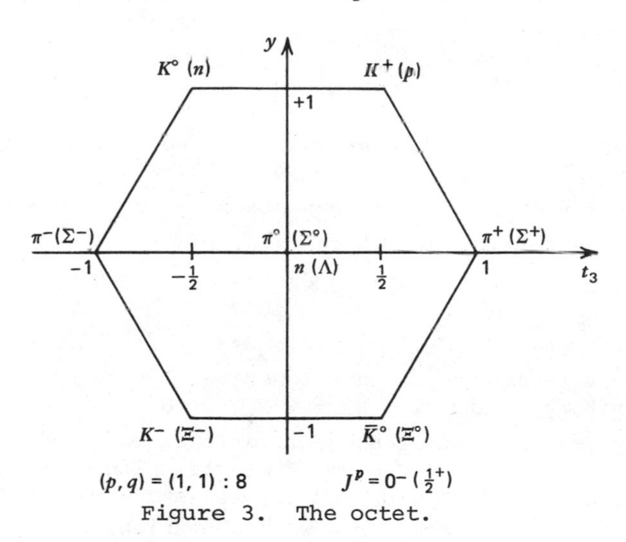

Figure 3. The octet.

Figure 3 shows the usual association of the octet with the 0^- mesons and the $1/2^+$ baryons (there are other octets with $J^P = 1^-$, 2^+). The octet is easily constructed using the aforementioned rules and the values given by our formulas $y_{max} = 1$ and $t_{max} = 1$.

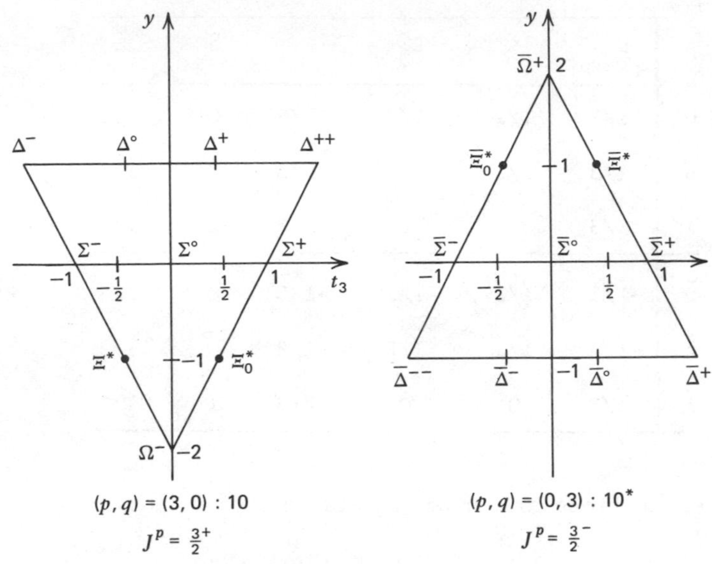

Figure 4. The decuplet.

The decuplets (Figure 4) are easily constructed after determining that $t_{max} = 3/2$, $y_{max} = 1$ for the 10 representation and $t_{max} = 3/2$, $y_{max} = 2$ for the 10* representation. The Gell-Mann-Nishijima relation is used to obtain the charges.

U SPIN

One may, alternatively, classify states by their eigenvalues of the operators U^2 and U_3: $U^2 = \frac{1}{2}(U_+U_- + U_-U_+) + (U_3)^2$. With U spin many consequences of SU(3) invariance can be determined without needing the SU(3) C-G coefficients. Since we know the commutators of T_3 with U_+, U_- and U_3, it is a simple calculation to show that the commutator of Q with U^2 vanishes. So one may use states that are simultaneously eigenstates of Q and U^2

$$U^2|U,U_3\rangle = U(U+1)|U, U_3\rangle$$

Furthermore, since

$$[U_\pm, Q] = [U_\pm, T_3] + \frac{1}{2}[U_\pm, Y] = \pm\frac{1}{2}U_\pm \mp \frac{1}{2}U_\pm = 0$$

all members of a U-spin multiplet have the same charge. Consider the baryon $\frac{1}{2}^+$ (meson 0^-) octet. The U-spin multiplets are

$$|U = \frac{1}{2}, U_3 = -\frac{1}{2}\rangle = |\frac{1}{2}, -\frac{1}{2}\rangle = |\Xi^-\rangle \quad (|k^-\rangle)$$

$$|\frac{1}{2}, \frac{1}{2}\rangle = |\Sigma^-\rangle \quad (|\pi^-\rangle)$$

$$|\frac{1}{2}, -\frac{1}{2}\rangle = |\Sigma^+\rangle \quad (|\pi^+\rangle)$$

$$|\frac{1}{2}, \frac{1}{2}\rangle = |p\rangle \quad (|k^+\rangle)$$

$$|0, 0\rangle = |\Lambda'\rangle \quad (|\eta'\rangle)$$

$$|1, -1\rangle = |\Xi^0\rangle \quad (|\bar{K}^0\rangle)$$

$$|1, 0\rangle = |\Sigma'\rangle \quad (|\pi'\rangle)$$

$$|1, 1\rangle = |\eta\rangle \quad (|K^0\rangle) \tag{29}$$

$|\Lambda\rangle$ and $|\Sigma_0\rangle$ are not eigenstates of U^2, but the linear combinations

$$|\Lambda'\rangle = -\frac{1}{2}|\Lambda\rangle - \frac{\sqrt{3}}{2}|\Sigma^0\rangle, \qquad |\Sigma'\rangle = \frac{\sqrt{3}}{2}|\Lambda\rangle - \frac{1}{2}|\Sigma^0\rangle$$

are:

$$(|\eta'\rangle = -\frac{1}{2}|\eta\rangle - \frac{\sqrt{3}}{2}|\pi^0\rangle), \qquad |\pi'\rangle = \frac{\sqrt{3}}{2}|\eta\rangle - \frac{1}{2}|\pi^0\rangle$$

One can use the commutation relations to verify these assignments.

For example,

$$U_3|n> = (\frac{3}{4} Y - T_{3/2})|n> = (\frac{3}{4} + \frac{1}{4})|n> = |n>$$

$$U^2|n> = \frac{1}{2} [U_+, U_-]| > + U_3{}^2|n>$$

$$= U_3(U_3 + 1)|n>$$

$$= 2|n> \text{ i.e., } U = 1 \qquad (30)$$

For the $(3/2)^+$ decuplet, the U-spin multiplets

$$|U = 0, U_3 = 0> \equiv |0,0> = |\Delta^{++}>$$

$$|\frac{1}{2}, -\frac{1}{2}> = |\Sigma^+>$$

$$|\frac{1}{2}, \frac{1}{2}> = |\Delta^+>$$

$$|1, -1> = |\Xi_0^*>$$

$$|1, 0> = |\Sigma^0>$$

$$|1, 1> = |\Delta^0>$$

$$|\frac{3}{2}, -\frac{3}{2}> = |\Omega^->$$

$$|\frac{3}{2}, -\frac{1}{2}> = |\Xi_-^*>$$

$$|\tfrac{3}{2}, \tfrac{1}{2}> \quad = \quad |\Sigma^->$$

$$|\tfrac{3}{2}, \tfrac{3}{2}> \quad = \quad |\Delta^-> \tag{31}$$

For example,

$$U_3|\Omega^-> \quad = \quad (\tfrac{3}{4}\,Y - T_{3/2})|\Omega^->$$

$$= \quad -\tfrac{3}{2}\,|\Omega^->$$

$$U^2|\Omega^-> \quad = \quad \{\tfrac{1}{2}\,(U_+U_- + U_-U_+) + U_3{}^2\}|\Omega^->$$

$$= \quad \tfrac{1}{2}\,[U_-, U_+]|\Omega^-> \;+ U_3{}^2|\Omega^->$$

$$= \quad U_3(U_3-1)\,|\Omega^->$$

$$= \quad \tfrac{3}{2} \times \tfrac{5}{2}\,|\Omega^-> \qquad \text{i.e., } U = \tfrac{3}{2} \tag{32}$$

For the fundamental three representation (quarks) the U-spin multiplets are

$$|U = 0, U_3 = 0> \;\equiv\; |0, 0> \quad = \quad |p>$$

$$|\tfrac{1}{2}, -\tfrac{1}{2}> \quad = \quad |\lambda>$$

$$|\tfrac{1}{2}, \tfrac{1}{2}> \quad = \quad |n> \tag{33}$$

GELL-MANN OKUBO MASS FORMULA

If one assumes that the mass operator is the sum of two terms, one which transforms like a U-spin scalar (U = 0) and the other which transforms like U = 1 (this is equivalent to the usual "octet enhancement" assumption) one obtains $M = a + bU_3$ for a given U-spin multiplet. For the ½+ octet,

$$M(\Xi^0) = a - b$$

$$M(\Sigma') = a = (3/4)M(\Lambda) + (1/4)M(\Sigma^0)$$

$$M(n) = a + b$$

so that $M(\Xi^0) + M(n) = M(\Sigma')$, which leads to the famous mass formula

$$\frac{M(\Xi^0) + M(n)}{2} = \frac{3M(\Lambda) + M(\Sigma^0)}{4} \quad (34)$$

For the 0^- octet the corresponding formula (using M^2 for mesons instead of M)

$$\frac{M^2(\bar{k}^0) + M^2(k^0)}{2} = M^2(k^0) = \frac{3\,M^2(\eta) + M^2(\pi^0)}{4} \quad (35)$$

For the $(3/2)^+$ decuplet,

$$M(\Omega^-) = a - (3/2)b$$

$$M(\Xi^*_-) = a - \tfrac{1}{2}b$$

$$M(\Sigma^-) = a + \tfrac{1}{2}b$$

$$M(\Delta^-) = a + (3/2)b \quad (36)$$

which leads to the "equal spacing rule"

$$M(\Omega^-) - M(\Xi^*) = M(\Xi^*_-) - M(\Sigma^-) = M(\Sigma^-) - M(\Delta^-) \quad (37)$$

PHOTOPRODUCTION

Consider the reactions

$$\left.\begin{array}{l}\gamma + p \to N + \pi^{+} \\ \gamma + p \to \Sigma' + k^{+}\end{array}\right\}$$

and

$$\left.\begin{array}{l}\gamma + p \to k^{o} + \Sigma^{+} \\ \gamma + p \to \pi' + p\end{array}\right\}$$

The photon γ is a U-spin scalar (U = 0). So,

$$|\gamma p> \;=\; |U = \tfrac{1}{2},\, U_3 = \tfrac{1}{2}> \;\equiv\; |\tfrac{1}{2}, \tfrac{1}{2}>$$

Using the ordinary SU(2) C-G coefficients to add U-spin

$$|\tfrac{3}{2}, \tfrac{1}{2}> \;=\; \sqrt{\tfrac{1}{3}}\,|1,1>\,|\tfrac{1}{2}, -\tfrac{1}{2}> + \sqrt{\tfrac{2}{3}}\,|1,0>|\tfrac{1}{2}, \tfrac{1}{2}>$$

$$|\tfrac{1}{2}, \tfrac{1}{2}> \;=\; \sqrt{\tfrac{2}{3}}\,|1,1>\,|\tfrac{1}{2}, -\tfrac{1}{2}> - \sqrt{\tfrac{1}{3}}\,|1,0>|\tfrac{1}{2}, \tfrac{1}{2}> \qquad (38)$$

So that

$$|1,1>\,|\tfrac{1}{2}, -\tfrac{1}{2}> = \sqrt{\tfrac{1}{3}}\,|\tfrac{3}{2}, \tfrac{1}{2}> + \sqrt{\tfrac{2}{3}}\,|\tfrac{1}{2}, \tfrac{1}{2}> = |N\,\pi^{+}>$$

$$(|k^{o}\Sigma^{+}>|$$

and

$$|1,0>\,|\tfrac{1}{2}, \tfrac{1}{2}> = \sqrt{\tfrac{2}{3}}\,|\tfrac{3}{2}, \tfrac{1}{2}> - \sqrt{\tfrac{1}{3}}\,|\tfrac{1}{2}, \tfrac{1}{2}>$$

$$= |\Sigma' k^{+}> \qquad (|\pi^{-}p>)$$

Therefore,

$$M(\gamma + p \to N + \pi^+) = \langle \gamma p|S|N\,\pi^+\rangle = \sqrt{\frac{2}{3}}\,\langle \tfrac{1}{2}, \tfrac{1}{2}|S|\tfrac{1}{2}, \tfrac{1}{2}\rangle$$

since the S matrix is a U-spin scalar, and

$$M(\gamma + p \to \Sigma' + k^+) = -\sqrt{\frac{1}{3}}\,\langle \tfrac{1}{2}, \tfrac{1}{2}\,|S|\,\tfrac{1}{2}, \tfrac{1}{2}\rangle$$

Using $|\Sigma'\rangle = \sqrt{\frac{3}{2}}\,|\Lambda\rangle - \frac{1}{2}\,|\Sigma^o\rangle$ we obtain

$$M(\gamma + p \to N + \pi^+) = -\sqrt{\frac{3}{2}}\,M(\gamma + p \to \Lambda + k^+)$$

$$+ \frac{1}{\sqrt{2}}\,M(\gamma + p \to \Sigma^o + k^+)$$

Similarly,

$$M(\gamma + p \to k^o + \Sigma^+) = -\sqrt{\frac{3}{2}}\,M(\gamma + p \to \eta + p)$$

$$+ \frac{1}{\sqrt{2}}\,M(\gamma + p \to \pi^o + p).$$

BIBLIOGRAPHY

Bethe, H. A. (1929) Annalen der Physik 3, 133.

Bouckaert, L. P., Smoluchowski, R., and Wigner, E. (1936) Physical Review 50, 58.

Brand, L. (1947) "Vector and Tensor Analysis," Wiley, New York.

Campbell, J. E. (1966) "Introductory Treatise on Lie's Theory of Finite Continuous Transformation Groups," Chelsea, New York.

Cotton, F. A. (1971) "Chemical Applications of Group Theory," Wiley-Interscience, New York.

DeShalit, A., and Talmi, I. (1963) "Nuclear Shell Theory," Academic, New York.

Gasiorowicz, S. (1966) "Elementary Particle Physics," Wiley, New York.

Gell-Mann, M., and Neeman, Y. (1964) "The Eightfold Way," Benjamin, New York.

Hamermesh, M. (1962) "Group Theory," Addison Wesley, Reading, Mass.

Herzberg, G. (1959) "Infrared and Raman Spectra," Van Nostrand, Princeton, N. J.

Lomont, J. S. (1959) "Applications of Finite Groups," Academic, New York.

Lichtenberg, D. B. (1970) "Unitary Symmetry and Elementary Particles," Academic, New York.

Lie, S. (1893) "Theorie der Transformationgruppen," F. Engel, Leipzig, Germany.

Littlewood, D. E. (1970) "A University Algebra," Dover, New York.

Meijer, P. H. (1964) (Ed.) "Group Theory and Solid State Physics," Gordon and Breach, New York.

Melvin, M. A. (1956) "Simplification in Finding Symmetry Adapted Eigenfunctions," Rev. Mod. Phys. 28, 18.

Messiah, A. (1962) "Quantum Mechanics," Interscience, New York.

Okubo, S. "Lectures on Unitary Symmetry," (Unpublished).

Schiff, L. I. (1968) "Quantum Mechanics," McGraw-Hill, New York.

Slater, J. C. (1972) "Symmetry and Energy Bands in Crystals," Dover, New York.

Tinkham, M. (1975) "Group Theory and Quantum Mechanics," McGraw-Hill, New York.

Wigner, E. P. (1959) "Group Theory and Its Application to the Quantum Mechanics of Atomic Spectra," Academic, New York.

APPENDIX A

SPHERICAL HARMONICS

The Legendre polynomials are defined as

$$P_\ell(\cos\,\theta) = \{1/2^\ell\ \ell!\}\ [d^\ell/d\ (\cos\,\theta)^\ell]\ (-\ \sin^2\,\theta)^\ell \qquad (1)$$

The (normalized) associated legendre polynomials are, with m > o,

$$(-)^m\{(2\ell+1/2)[(\ell-m)!/(\ell+m)!]\}^{\frac{1}{2}}\ \sin^m\,\theta\ (d^m/d(\cos\,\theta)^m)$$

$$\times\ (P_\ell(\cos\,\theta)) \qquad (2)$$

$$P_\ell^{-m} \;=\; (-)^m P_\ell^m \qquad (3)$$

The normalized spherical harmonics are defined as

$$Y_\ell^m\ (\hat{r}) \equiv Y_\ell^m\ (\theta,\psi) \;=$$

$$(-)^m\ \{[(2\ell+1)/2]\ (\ell-m)!/(\ell+m)!\}^{\frac{1}{2}}\ [\sin^m\,\theta(d^m/d(\cos\,\theta)^m$$

$$\times\ (P_\ell(\cos\,\theta)]\ \cdot\ [\tfrac{1}{2}\pi]^{\frac{1}{2}}\ \exp(im\psi) \qquad (4)$$

$$= \{\tfrac{1}{2}\pi\}^{\frac{1}{2}}\ P_\ell^m(\cos\,\theta)\ \exp(im\psi) \qquad (5)$$

$$Y_\ell^{m*} = (-)^m Y_\ell^{-m} \tag{6}$$

The first few spherical harmonics are

$$Y_o^o = \{1/4\pi\}^{\frac{1}{2}}$$

$$Y_1^o = \{3/4\pi\}^{\frac{1}{2}} \cos(\theta) = \{3/4\pi\}^{\frac{1}{2}} Z/r$$

$$Y_1^{\pm} = \mp \{3/8\pi\}^{\frac{1}{2}} \sin\theta \exp \pm i\psi = \mp \{3/8\pi\}^{\frac{1}{2}} (x \pm iy)/r$$

$$Y_2^o = \{5/16\pi\}^{\frac{1}{2}} (3 \cos^2\theta - 1) = \{5/16\pi\}^{\frac{1}{2}} (3Z^2 - r^2)/r^2$$

$$Y_2^{\pm 1} = \mp\{15/8\pi\}^{\frac{1}{2}} \sin\theta \cos\theta \exp(\pm i\psi) = \mp \{15/8\pi\}^{\frac{1}{2}} \times (X \pm iy) Z/r^2$$

$$Y_2^{\pm 2} = \{15/32\pi\}^{\frac{1}{2}} \sin^2\theta \exp(\pm 2i\psi) = \{15/32\pi\}^{\frac{1}{2}} \cdot (x \pm iy)^2/r^2 \tag{7}$$

The ortho-normal property of the Y_ℓ^m is expressed as

$$< \ell m | \ell' m' > = (Y_\ell^m, Y_{\ell'}^{m'}) \equiv \int_o^{2\pi} \int_o^{\pi} Y_\ell^{m*} Y_{\ell'}^{m'} d\Omega = \delta_{\ell\ell'} \delta_{mm'}$$

$$d\Omega = \sin\theta \, d\theta \, d\psi$$

$$\delta_{ab} = 1 \text{ when } a = b$$

$$= o \text{ otherwise} \tag{8}$$

The Y_ℓ^m functions satisfy the differential equation

$$[(1/\sin\theta)\partial/\partial\theta\ (\sin\theta\ \partial/\partial\theta) + (1/\sin^2\theta)\ \partial^2/\partial\psi^2$$

$$+ \ell(\ell+1)]\ Y_\ell^m = 0 \tag{9}$$

or

$$\{-\vec{L}^2 + \ell(\ell+1)\}\ Y_\ell^m = 0$$

$$\vec{L} = -i\ \vec{r} \times \vec{\nabla} \tag{10}$$

The parity of the spherical harmonics is given by

$$\underset{\sim}{\Pi}\ Y_\ell^m\ (\hat{r}) = Y_\ell^m\ (-\hat{r}) = Y_\ell^m(\pi-\theta,\ \pi+\psi) = (-)^\ell\ Y_\ell^m\ (\theta,\psi) \tag{11}$$

We now write down a series of useful relations

$$(\partial/\partial\theta)\ Y_\ell^m =$$

$$\exp(-i\psi)\ \{[(\ell-m)(\ell+m+1)]/4\}^{\frac{1}{2}}\ Y_\ell^{m+1} - \exp(i\psi)$$

$$\times \left\{ \frac{(\ell+m)(\ell-m+1)}{4} \right\}^{\frac{1}{2}} Y_\ell^{m-1} \tag{12}$$

$$\text{Cos}\ \theta\ Y_\ell^m = [(\ell-m+1)(\ell+m+1)/(2\ell+1)(2\ell+3)]^{\frac{1}{2}}\ Y_{\ell+1}^m$$

$$+ \left[\frac{(\ell+m)(\ell-m)}{(2\ell-1)(2\ell+1)} \right]^{\frac{1}{2}} Y_\ell^m{-1} \tag{13}$$

$$\sin\theta\, Y_\ell^m = \left\{ -[(\ell+m+1)(\ell+m+2)/(2\ell+1)(2\ell+3)]^{\frac{1}{2}}\, Y_{\ell+1}^{m+1} + [(\ell-m-1)(\ell-m)/(2\ell-1)(2\ell+1)]^{\frac{1}{2}}\, Y_{\ell-1}^{m+1} \right\} \exp(-i\psi) \tag{14}$$

also =

$$\{(\ell-m+1)(\ell-m+2)/(2\ell+1)(2\ell+3)\}^{\frac{1}{2}}\, Y_{\ell+1}^{m-1} \exp(+i\psi)$$

$$- (\ell+m-1)(\ell+m)/(2\ell-1)(2\ell+1)\}^{\frac{1}{2}}\, Y_{\ell-1}^{m-1} \exp(+i\psi) \tag{15}$$

because of the identity

$$[(\ell-m-1)(\ell-m)/(2\ell-1)(2\ell+1)]^{\frac{1}{2}}\, Y_{\ell-1}^{m+1} \exp(-i\psi)$$

$$+ [(\ell+m-1)(\ell+m)/(2\ell-1)(2\ell+1)]^{\frac{1}{2}}\, Y_{\ell-1}^{m-1} \exp(i\psi) \tag{16}$$

$$= [(\ell+m+1)(\ell+m+2)/(2\ell+1)(2\ell+3)]^{\frac{1}{2}}\, Y_{\ell+1}^{m+1} \exp(-i\psi)$$

$$+ [(\ell-m+1)(\ell-m+2)/(2\ell+1)(2\ell+3)]^{\frac{1}{2}}\, Y_{\ell+1}^{m-1} \exp(i\psi) \tag{17}$$

$$\sin\theta\, \partial/\partial\theta\, Y_\ell^m = \ell\{(\ell+m+1)(\ell-m+1)/(2\ell+1)(2\ell+3)\}^{\frac{1}{2}}\, Y_{\ell+1}^m$$

$$- (\ell+1)\, \{(\ell-m)(\ell+m)/(2\ell-1)(2\ell+1)\}^{\frac{1}{2}}\, Y_{\ell-1}^m \tag{18}$$

$$\cos\theta\, \partial/\partial\theta\, Y_\ell^m = (\ell+1)/2\, \{[(\ell-m-1)(\ell-m)/(2\ell-1)(2\ell+1)]^{\frac{1}{2}}$$
$$\times\, Y_{\ell-1}^{m+1} \exp(-i\psi)$$

$$- [(\ell+m-1)(\ell+m)/(2\ell-1)(2\ell+1)]^{\frac{1}{2}} Y_{\ell-1}^{m-1} \exp(i\psi)\}$$

$$+ \ell/2 \{[(\ell+m+1)(\ell+m+2)/(2\ell+1)(2\ell+3)]^{\frac{1}{2}} Y_{\ell+1}^{m+1} \exp(-i\psi)$$

$$- [(\ell-m+1)(\ell-m+2)/(2\ell+1)(2\ell+3)]^{\frac{1}{2}} Y_{\ell+1}^{m-1} \exp(i\psi)\} \qquad (19)$$

also =

$$\tfrac{1}{2}[(\ell-m)(\ell+m+1)]^{\frac{1}{2}} \{[(\ell-m)(\ell+m+2)/(2\ell+1)(2\ell+3)]^{\frac{1}{2}} Y_{\ell+1}^{m+1} \exp(-i\psi)$$

$$+ [(\ell-m-1)(\ell+m+1)/(2\ell-1)(2\ell+1)]^{\frac{1}{2}} Y_{\ell-1}^{m+1} \exp -i\psi\} \qquad (20)$$

$$- \tfrac{1}{2}[(\ell+m)(\ell-m+1)]^{\frac{1}{2}} \{[(\ell-m+2)(\ell+m)/(2\ell+1)(2\ell+3)]^{\frac{1}{2}}$$

$$\cdot Y_{\ell+1}^{m-1} \exp i\psi + [(\ell-m+1)(\ell+m-1)/(2\ell-1)(2\ell+1)]^{\frac{1}{2}}$$

$$\cdot Y_{\ell-1}^{m-1} \exp i\psi\} \qquad (21)$$

$$[i/\sin\theta]\ \partial/\partial\psi\ Y_{\ell}^{m} = (2\ell+1)/2\{[(\ell+m+1)(\ell+m+2)/(2\ell+1)(2\ell+3)]^{\frac{1}{2}}$$

$$\cdot\ Y_{\ell+1}^{m+1} \exp(-i\psi)\} + [(\ell-m+1)(\ell-m+2)/(2\ell+1)(2\ell+3)]$$

$$\cdot\ Y_{\ell+1}^{m-1} \exp(i\psi)\} \qquad (22)$$

also =

$$(2\ell+1)/2\ \{[(\ell-m)(\ell-m-1)/(2\ell-1)(2\ell+1)]^{\frac{1}{2}} Y_{\ell-1}^{m+1} \exp(-i\psi)$$

$$+ \left[(\ell+m-1)(\ell+m)/(2\ell-1)(2\ell+1)\right]^{\frac{1}{2}} Y_{\ell-1}^{m-1} \exp(i\psi)\} \quad (23)$$

$$([\cos\theta\ \partial/\partial\theta] + (i/\sin\theta)\ \partial/\partial\psi)\ Y_{\ell}^{m} = \ell[(\ell+m+1)(\ell+m+2)/$$

$$(2\ell+1)(2\ell+3)]^{\frac{1}{2}} Y_{\ell+1}^{m+1} \exp(-i\psi)$$

$$+ (\ell+1)\ [(\ell-m)(\ell-m-1)/(2\ell-1)(2\ell+1)]^{\frac{1}{2}}$$

$$\cdot Y_{\ell-1}^{m+1} \exp(-i\psi) \quad (24)$$

$$([\cos\theta\ \partial/\partial\theta] - (i/\sin\theta)\ \partial/\partial\psi)\ Y_{\ell}^{m} = -(\ell+1)$$

$$\times \left[\frac{(\ell+m-1)(\ell+m)}{(2\ell-1)(2\ell+1)}\right]^{\frac{1}{2}} Y_{\ell-1}^{m-1} \exp(i\psi)$$

$$- \ell[(\ell-m+1)(\ell-m+2)/(2\ell+1)(2\ell+3)]^{\frac{1}{2}} Y_{\ell+1}^{m-1} \exp(i\psi) \quad (25)$$

$$\underset{\sim}{L}_x = i(\sin\psi\ \partial/\partial\theta + \cot\theta\ \cos\psi\ \partial/\partial\psi)$$

$$\underset{\sim}{L}_y = i(-\cos\psi\ \partial/\partial\theta + \cot\theta\ \sin\psi\ \partial/\partial\psi) \quad (26)$$

$$\underset{\sim}{L}_z = -1\ \partial/\partial\psi$$

$$\underset{\sim}{L}^2 = \underset{\sim}{L}_x \underset{\sim}{L}_x + \underset{\sim}{L}_y \underset{\sim}{L}_y + \underset{\sim}{L}_z \underset{\sim}{L}_z \quad (27)$$

$$= -\{(1/\sin\theta)\ \partial/\partial\theta(\sin\theta\ \partial/\partial\theta) + (1/\sin^2\theta)\ \partial^2/\partial\psi^2\} \quad (28)$$

$$\underset{\sim}{L}^2 \; Y_\ell^m = \ell(\ell+1) \; Y_\ell^m$$

$$\underset{\sim}{L}_z \; Y_\ell^m = mY_\ell^m \tag{29}$$

$$\underset{\sim}{L}_+ \; Y_\ell^m \equiv (\underset{\sim}{L}_x + i\underset{\sim}{L}_y) \; Y_\ell^m = \{(\ell-m)(\ell+m+1)\}^{\frac{1}{2}} \; Y_\ell^{m+1}$$

$$\underset{\sim}{L}_- \; Y_\ell^m \equiv (\underset{\sim}{L}_x - i\underset{\sim}{L}_y) \; Y_\ell^m = \{(\ell+m)(\ell-m+1)\}^{\frac{1}{2}} \; Y_\ell^{m-1} \tag{30}$$

$\underset{\sim}{L}_+$, $\underset{\sim}{L}_-$ step up and down, respectively, the m value and hence are the ladder operators.

The following properties of the spherical harmonics are frequently used:

$$\sum_{\ell=0}^{\infty} \sum_{m=-\ell}^{\infty} Y_\ell^{m*} \; (\theta_1\psi_1) \; Y_\ell^m \; (\theta_2\psi_2) = \delta(\psi_1-\psi_2) \; \delta(\cos\theta_1-\cos\theta_2)$$

(closure property) (31)

$$\sum_{m=-\ell}^{\ell} |Y_\ell^m(\theta,\psi)|^2 = (2\ell+1)/4\pi \quad \text{(sum rule)} \tag{32}$$

$$P_\ell \; (\cos\Upsilon) = (4\pi/2\ell+1) \sum_{m=-\ell}^{\ell} Y_\ell^{m*} \; (\theta_1\psi_1) \; Y_\ell^m(\theta_2\psi_2) \tag{33}$$

where $\cos\Upsilon = \cos\theta_1 \cos\theta_2 + \sin\theta_1 \sin\theta_2 \cos(\psi_1 - \psi_2)$ (addition theorem)

$$Y_L^M(\theta\psi) \; Y_\ell^m(\theta\psi) = \Sigma_k Y_k^{M+m}(\theta\psi) \; C_{M,m,M+m}^{L\ell k} \; C_{ooo}^{L\ell k} \left\{\frac{(2L+1)(2\ell+1)}{4\pi(2k+1)}\right\}^{\frac{1}{2}} \tag{34}$$

where the Clebsch-Gordan coefficients are, in customary notation,

$$C^{\ell_1 \ell_2 L}_{m_1 m_2 M} \equiv C(\ell_1 \ell_2 L; m_1 m_2) \equiv C(\ell_1 \ell_2 L; m_1 m_2 M)$$

$$\int Y^{m_3 *}_{\ell_3} Y^{m_2}_{\ell_2} Y^{m_1}_{\ell_1} d\Omega = \{(2\ell_1 + 1)(2\ell_2 + 1)/4\pi(2\ell_3 + 1)\}^{\frac{1}{2}}$$

$$\times C^{\ell_1 \ell_2 \ell_3}_{m_1 m_2 m_3} C^{\ell_1 \ell_2 \ell_3}_{o\ o\ o} \qquad (35)$$

where the Cs are Clebsch-Gordan coefficients.

$$Y^{M*}_{L}(\beta,\alpha) = \{(2\ell + 1)/4\pi\}^{\frac{1}{2}} D^{(L)}_{Mo}(\alpha,\beta,o) \qquad (36)$$

where the $D^{(L)}_{m'm}(\alpha,\beta,\gamma)$ are the (2L+1) dimensional irreducible representations of the three-dimensional rotation group

$$1/|\vec{r}_1 - \vec{r}_2| = 4\pi \sum_{\ell=o}^{\infty} \sum_{m=-\ell}^{\ell} (1/2\ell+1)(r_<^{\ell}/r_>^{\ell+1}) Y^{m*}_{\ell}(\theta_2\psi_2) Y^{m}_{\ell}(\theta_1\psi_1)$$

$$r_</r_> = r_1/r_2 \text{ if } r_2 > r_1, \quad = r_2/r_1 \text{ if } r_2 < r_1 \qquad (37)$$

$$\exp(i\vec{k}\cdot\vec{r}) = 4\pi \sum_{\ell=o}^{\infty} i^{\ell} j_{\ell}(kr) \sum_{m=-\ell}^{\ell} Y^{m*}_{\ell}(\theta\psi) Y^{m}_{\ell}(\theta_k\psi_k) \qquad (38)$$

where θ_k, ψ_k (with subscript k) are the polar angles of the vector $\vec{k}$, while θ, ψ are the polar angles of the vector $\vec{r}$."
$j_{\ell}(kr)$ is defined as $(\pi/2kr)^{\frac{1}{2}} J_{\ell+\frac{1}{2}}(kr)$ where $J_{\ell+\frac{1}{2}}$ is a Bessel function of order $\ell + \frac{1}{2}$. The spherical harmonics satisfy the following generating function:

$$\{-(x-iy) + 2zt + (x+iy)t^2\}^{\ell} = 2^{\ell} \ell! \sum_{m=-\ell}^{\ell} (-)^m t^{\ell+m}/(\ell+m)!$$

$$\cdot[\{(2\ell+1)/4\pi\}\ (\ell-|m|)!/(\ell+|m|)!]^{-\frac{1}{2}}\ r^{\ell}Y_{\ell}^{m} \tag{39}$$

In the Pauli theory of the spinning electron and the relativistic Dirac theory the following spinors are used:

$$\chi_{x}^{m} = \begin{pmatrix} \{(\ell+\mu+\frac{1}{2})/2\ell+1\}^{\frac{1}{2}}\ Y_{\ell}^{\mu-\frac{1}{2}} \\ \{(\ell-\mu+\frac{1}{2})/(2\ell+1)\}^{\frac{1}{2}}\ Y_{\ell}^{\mu+\frac{1}{2}} \end{pmatrix} \qquad J = \ell + \tfrac{1}{2} \tag{40}$$

x here is the Dirac quantum number

$$x = -\ell - 1 \qquad j = \ell + \tfrac{1}{2}$$

$$= +\ell \qquad = \ell - \tfrac{1}{2}$$

$|x| = j + \frac{1}{2}$ always

$$\chi_{x}^{\mu} = \begin{pmatrix} -\{(\ell-\mu+\frac{1}{2})/2\ell+1\}^{\frac{1}{2}}\ Y_{\ell}^{\mu-\frac{1}{2}} \\ \{(\ell+\mu+\frac{1}{2})/2\ell+1\}^{\frac{1}{2}}\ Y_{\ell}^{\mu+\frac{1}{2}} \end{pmatrix} \qquad j = \ell - \tfrac{1}{2} \tag{41}$$

χ_{-x}^{μ} is defined as follows

$$\chi_{-x}^{\mu} = \begin{pmatrix} -\{(\ell-\mu+3/2)/2\ell+3\}^{\frac{1}{2}}\ Y_{\ell+1}^{\mu-\frac{1}{2}} \\ \{(\ell+\mu+3/2)/2\ell+3\}^{\frac{1}{2}}\ Y_{\ell+1}^{\mu+\frac{1}{2}} \end{pmatrix} \qquad j = \ell + \tfrac{1}{2} \tag{42}$$

$$= \begin{pmatrix} \{(\ell+\mu-\frac{1}{2})/2\ell-1\}^{\frac{1}{2}} \; Y^{\mu-\frac{1}{2}}_{\ell-1} \\ \{(\ell-\mu-\frac{1}{2})/2\ell-1\}^{\frac{1}{2}} \; Y^{\mu+\frac{1}{2}}_{\ell-1} \end{pmatrix} \qquad j = \ell - \tfrac{1}{2} \tag{43}$$

$$\ell(x) = \ell$$

$$\ell(-x) = \ell + 1, \qquad j = \ell + \tfrac{1}{2}$$

$$= \ell - 1, \qquad j = \ell - \tfrac{1}{2}$$

and the general definition of these spinors is

$$\chi^{\mu}_{x} = \sum_{T} C(\ell(x)\tfrac{1}{2}\, j;\;\; \mu - T, T, \mu) \;\; Y^{\mu-T}_{\ell(x)} \; \chi^{T}_{\frac{1}{2}} \tag{44}$$

where $\chi^{T}_{\frac{1}{2}}$ are the spinfunctions $\chi^{\frac{1}{2}}_{\frac{1}{2}} = \alpha = \begin{pmatrix} 1 \\ 0 \end{pmatrix}$

$$\chi^{-\frac{1}{2}}_{\frac{1}{2}} = \beta = \begin{pmatrix} 0 \\ 1 \end{pmatrix}$$

Two tables of Clebsch-Gordan coefficients of use here are appended. The χ^{μ}_{x} satisfy the equations

$$(\vec{\sigma}\cdot\vec{L} + 1)\; \chi^{\mu}_{x} = -\, x\chi^{\mu}_{x}$$

$$\frac{\vec{\sigma}\cdot\vec{r}}{r}\, \chi^{\mu}_{x} \equiv \vec{\sigma} \cdot \hat{r}\, \chi^{\mu}_{x} = -\, \chi^{\mu}_{-x} \tag{45}$$

$\hat{r}$ is the unit vector $\vec{r}/r$. Before going to vector spherical harmonics or irreducible tensors, let us introduce the relations pertaining to a spherical basis:

$$\vec{A} = A_x\, \hat{x} + A_y\, \hat{y} + A_z\, \hat{z} \tag{46}$$

The unit vectors in a spherical basis are defined as

$$\hat{\xi}_1 = -\frac{1}{\sqrt{2}} (\hat{x} + i\,\hat{y})$$

$$\hat{\xi}_{-1} = +\frac{1}{\sqrt{2}} (\hat{x} - i\,\hat{y})$$

$$\hat{\xi}_o = \hat{z} \tag{47}$$

The components of $\vec{A}$ in the spherical basis are

$$A_1 = -\frac{1}{\sqrt{2}} (A_x + i\,A_y)$$

$$A_{-1} = +\frac{1}{\sqrt{2}} (A_x - i\,A_y)$$

$$A_o = A_z \tag{48}$$

The vector is now written as

$$\vec{A} = -A_1\,\hat{\xi}_{-1} - A_{-1}\,\hat{\xi}_1 + A_o\hat{\xi}_o$$

$$= \sum_{\mu} (-)^{\mu} A_{\mu}\,\hat{\xi}_{-\mu} \tag{49}$$

The components of $\vec{\nabla}$ and $\vec{\nabla}$x are $(\vec{\nabla})_o = \text{Cos}\,\theta\,\partial/\partial r - (1/r)\sin\theta\,\partial/\partial\theta$

$$(\vec{\nabla})_1 = -\frac{1}{\sqrt{2}} \exp i\psi\,\{\sin\theta\,\partial/\partial r + (1/r)$$

$$\times\,\text{Cos}\,\theta\,\partial/\partial\theta + i/(r\sin\theta)\,\partial/\partial\psi\}$$

$$(\vec{\nabla})_{-1} = + \frac{1}{\sqrt{2}} \exp -i\psi \, \{\sin\theta\, \partial/\partial r + (1/r)$$

$$\times \; \mathrm{Cos}\,\theta\, \partial/\partial\theta - i/(r \sin\theta)\, \partial/\partial\psi\}$$

$$(\vec{\nabla} \times \vec{A})_1 = i\{[-(1/r)\sin\theta\, \partial/\partial\theta + \cos\theta\, \partial/\partial r]\, A_1$$

$$+ \frac{1}{\sqrt{2}} \exp i\psi \, [\sin\theta\, \partial/\partial r + (1/r)\cos\theta\, \partial/\partial\theta$$

$$+ \; i/(r\sin\theta)\, \partial/\partial\psi]\, A_o\} \qquad (50)$$

$$(\vec{\nabla} \times \vec{A})_{-1} = i \left(\exp(-\,i\psi)\frac{1}{\sqrt{2}} [\sin\theta\, \partial/\partial r + (1/r) \right.$$

$$\times \; \mathrm{Cos}\,\theta\, \partial/\partial\theta - (i/r\sin\theta)\, \partial/\partial\psi]\, A_o$$

$$\left. - \; [\mathrm{Cos}\,\theta\, \partial/\partial r - \frac{1}{r}\sin\theta\, \partial/\partial\theta]\, A_{-1}\} \right) \qquad (51)$$

$$(\vec{\nabla} \times \vec{A})_o = i \exp(-i\psi)\frac{1}{\sqrt{2}} \{\sin\theta\, \partial/\partial r + (1/r)$$

$$\times \; \cos\, \partial/\partial\theta - i/(r\sin\theta)\partial/\partial\psi\}\, A_1$$

$$+ \; i \exp(i\psi)\frac{1}{\sqrt{2}} \{\sin\theta\, \partial/\partial r + (1/r)$$

$$\times \; \cos\theta\partial/\partial\theta + i/(r\sin\theta)\partial/\partial\psi\}\, A_{-1}$$

In general,

$$(\vec{A} \times \vec{B})_1 = i\,(A_o B_1 - A_1 B_o)$$

$$(\vec{A} \times \vec{B})_{-1} = i\,(A_{-1}B_{o} - A_{o}B_{-1})$$

$$(\vec{A} \times \vec{B})_{o} = i\,(A_{-1}B_{1} - A_{+1}B_{-1}) \tag{52}$$

Written in matrix form, the vector spherical harmonics are

$$\left[A = \begin{pmatrix} -A_{-1} \\ A_{o} \\ -A_{1} \end{pmatrix} \right]$$

$$\vec{Y}^{M}_{JJ} = \begin{pmatrix} -[(J+M)(J-M+1)/2J(J+1)]^{\frac{1}{2}} \quad Y^{M-1}_{J} \\ M/\sqrt{J(J+1)} \quad Y^{M}_{J} \\ [(J-M)(J+M+1)/2J(J+1)]^{\frac{1}{2}} \quad Y^{M+1}_{J} \end{pmatrix} \tag{53}$$

$$\vec{Y}^{M}_{J,J-1} = \begin{pmatrix} [(J+M-1)(J+M)/2J(2J-1)]^{\frac{1}{2}} \quad Y^{M-1}_{J-1} \\ [(J-M)(J+M)/J(2J-1)]^{\frac{1}{2}} \quad Y^{M}_{J-1} \\ [(J-M-1)(J-M)/2J(2J-1)]^{\frac{1}{2}} \quad Y^{M+1}_{J-1} \end{pmatrix} \tag{54}$$

$$\vec{Y}^{M}_{J,J+1} = \begin{pmatrix} [(J-M+1)(J-M+2)/2(J+1)(2J+3)]^{\frac{1}{2}} \quad Y^{M-1}_{J+1} \\ -[(J-M+1)(J+M+1)/(J+1)(2J+3)]^{\frac{1}{2}} \quad Y^{M}_{J+1} \\ [(J+M+1)(J+M+2)/2(J+1)(2J+3)]^{\frac{1}{2}} \quad Y^{M+1}_{J+1} \end{pmatrix} \tag{55}$$

In other words, the unit vectors are chosen

$$\hat{\xi}_1 = \begin{pmatrix} 1 \\ 0 \\ 0 \end{pmatrix}; \quad \hat{\xi}_0 = \begin{pmatrix} 0 \\ 1 \\ 0 \end{pmatrix}; \text{ and } \hat{\xi}_{-1} = \begin{pmatrix} 0 \\ 0 \\ 1 \end{pmatrix} \tag{56}$$

and, for instance,

$$(\vec{Y}^M_{JJ})_{-1} = [(J+M)(J-M+1)/2J(J+1)]^{\frac{1}{2}} Y^{M-1}_J(\theta,\psi) \tag{57}$$

$$\vec{Y}^M_{JJ} = \underset{\sim}{\vec{L}} (Y^M_J/\sqrt{J(J+1)}) ;$$

$$\vec{Y}^M_{JL} = \sum_T C(L1J;M-T,T,M)\, Y^{M-T}_L\, \hat{\xi}_T \tag{58}$$

The vector spherical harmonics satisfy the eigenvalue equations

$$J^2 \vec{Y}^M_{JL} = J(J+1) \vec{Y}^M_{JL}$$

$$J_z \vec{Y}^M_{JL} = M \vec{Y}^M_{JL}$$

$$\vec{J} = \vec{L} + \vec{S} \tag{59}$$

$$S_x = \frac{1}{\sqrt{2}} \begin{pmatrix} 0 & 1 & 0 \\ 1 & 0 & 1 \\ 0 & 1 & 0 \end{pmatrix} \qquad S_y = \frac{1}{\sqrt{2}} \begin{pmatrix} 0 & -i & 0 \\ i & 0 & -i \\ 0 & i & 0 \end{pmatrix}$$

$$S_z = \begin{pmatrix} 1 & 0 & 0 \\ 0 & 0 & 0 \\ 0 & 0 & -1 \end{pmatrix} \quad S^2 = S_x^2 + S_y^2 + S_z^2 = \begin{pmatrix} 2 & 0 & 0 \\ 0 & 2 & 0 \\ 0 & 0 & 2 \end{pmatrix}$$

$$S_1 = \begin{pmatrix} 0 & -1 & 0 \\ 0 & 0 & -1 \\ 0 & 0 & 0 \end{pmatrix} \quad S_{-1} = \begin{pmatrix} 0 & 0 & 0 \\ 1 & 0 & 0 \\ 0 & 1 & 0 \end{pmatrix} \quad S_o = \begin{pmatrix} 1 & 0 & 0 \\ 0 & 0 & 0 \\ 0 & 0 & -1 \end{pmatrix}$$

$$S_o \, \hat{\xi}_\mu = \mu \, \hat{\xi}_\mu$$

$$S^2 \, \hat{\xi}_\mu = 2 \, \hat{\xi}_\mu \tag{60}$$

If f(r) is an arbitrary (regular) function of r, we have

$$\vec{\nabla} \cdot f \, \vec{Y}^M_{JJ} = 0$$

$$\vec{\nabla} \cdot f \, \vec{Y}^M_{JJ-1} = -[J/2J+1]^{\frac{1}{2}} \{(J-1) \, f/r - df/dr\} \, Y^M_J$$

$$\vec{\nabla} \cdot f \, \vec{Y}^M_{JJ+1} = -[(J+1)/2J+1]^{\frac{1}{2}} \{(J+2) f/r + df/dr\} \, Y^M_J \tag{61}$$

$$-i \, \vec{\nabla} \times f \, \vec{Y}^M_{JJ} = [(J+1)/2J+1]^{\frac{1}{2}} \{(J+1) f/r + df/dr\} \, \vec{Y}^M_{JJ-1}$$

$$+ \; [J/2J+1]^{\frac{1}{2}} \{-Jf/r + df/dr\} \, \vec{Y}^M_{JJ+1} \tag{62}$$

$$-i \, \vec{\nabla} \times f \, \vec{Y}^M_{JJ-1} = -[(J+1)/2J+1]^{\frac{1}{2}} \{(J-1) f/r - df/dr\} \, \vec{Y}^M_{JJ}$$

$$-i \, \vec{\nabla} \times f \, \vec{Y}^M_{JJ+1} = [J/2J+1]^{\frac{1}{2}} \{(J+2) f/r + df/dr\} \, \vec{Y}^M_{JJ}$$

$$(\vec{Y}^M_{JL}, \vec{Y}^{M'}_{J'L'}) = \delta_{JJ'}, \, \delta_{LL'}, \, \delta_{MM'} \tag{63}$$

The $\vec{Y}^M_{JJ+1}$ and $\vec{Y}^M_{JJ-1}$ belong to the same parity $(-)^{J+1}$ which is opposite to that of $\vec{Y}^M_{JJ}$. If we chose $f_\ell = j_\ell$, these satisfy the following contiguous relations:

$$(d/d\rho)\ f_j(\rho) = f_{J-1}(\rho) - ((J+1)/\rho) f_J(\rho)$$

$$= (J/\rho) f_J - f_{J+1}$$

$$= (J/2J+1)\ f_{J-1} - ((J+1)/2J+1) f_{J+1}$$

$$f_{J-1} + f_{J+1} = [(2J+1)/\rho] f_J \tag{64}$$

We now introduce the Hansen functions

$$\vec{M}_{JM} = -f_L\ \vec{Y}^M_{JJ}$$

$$\vec{N}_{JM} = (+i/k)\vec{\nabla} \times \vec{M}_{JM}$$

$$= [(J+1)/2J+1]^{\frac{1}{2}}\ f_{J-1}\ \vec{Y}^M_{JJ-1} - [J/(2J+1)]^{\frac{1}{2}}\ f_{J+1}\ \vec{Y}^M_{JJ+1} \tag{65}$$

$$\vec{L}_{JM} = (1/k)\ \nabla(f_L Y^M_L)$$

$$= [J/2J+1]^{\frac{1}{2}}\ f_{J-1}\ \vec{Y}^M_{JJ-1} + [(J+1)/2J+1]^{\frac{1}{2}}\ f_{J+1}\ \vec{Y}^M_{JJ+1} \tag{66}$$

we then have the following relations:

$$\vec{\nabla} \cdot \vec{M}_{JM} = 0$$

$$\vec{\nabla} \cdot \vec{N}_{JM} = 0$$

$$\vec{\nabla} \times \vec{L}_{JM} = 0$$

$$\vec{\nabla} \cdot \vec{L}_{JM} = -k\, f_J\, Y_J^M$$

$$\vec{\nabla} \times \vec{M}_{JM} = -i\, k\, \vec{N}_{JM}$$

$$\vec{\nabla} \times \vec{N}_{JM} = +i\, k\, \vec{M}_{JM} \qquad (67)$$

The following relations involving $\vec{Y}^M_{JL}$ s are of use in electrodynamics:

$$\vec{Y}^M_{JJ-1} = -[1/J(2J+1)]^{\frac{1}{2}} \{-J\, \hat{r} + i\, \hat{r} \times \vec{L}\}\, Y_J^M$$

$$\vec{Y}^M_{JJ+1} = -[1/(J+1)(2J+1)]^{\frac{1}{2}} \{(J+1)\, \hat{r} + i\, \hat{r} \times \vec{L}\}\, Y_J^M$$

$$\hat{r}\, Y_J^M = -[(J+1)/2J+1]^{\frac{1}{2}}\, \vec{Y}^M_{JJ+1} + [J/2J+1]^{\frac{1}{2}}\, \vec{Y}^M_{JJ-1}$$

$$= -\sum_L C(J1L;000)\, \vec{Y}^M_{JL} \qquad (68)$$

$$\vec{\nabla} f(r)\, \vec{Y}^M_J = -\{(J+1)/2J+1\}^{\frac{1}{2}} (df/dr - (J/r)f)\, \vec{Y}^M_{JJ+1}$$

$$+ (J/J+1)^{\frac{1}{2}} \{df/dr + (J+1)f/r\}\, \vec{Y}^M_{JJ-1} \qquad (69)$$

$$\vec{Y}^{M*}_{JL} \cdot \vec{Y}^{M'}_{J'L'} =$$

$$\sum_{K=0}^{\infty} \sum_{T=-1}^{1} (-)^{M-T} \{(2L+1)(2L'+1)/4\pi(2k+1)\}^{\frac{1}{2}} C(L1J;M-T,T)\, x$$

$$x C(L'1J';M'-T,T)\, C(LL'K;-M+T,M'-T)\, C(LL'K;00)\, Y_K^{M'-M}$$

$$\vec{Y}^{M}_{JL} \times \vec{Y}^{M'}_{J'L'} = i\sqrt{2} \sum_{\mu\lambda\nu} \left[(2L+1)(2L'+1)/4\pi(2\nu+1)\right]^{\frac{1}{2}} \times$$

$$\times\ C(LL'\nu;00)\ C(111;\mu-\lambda,\lambda)\ C(L'1J';M'-\lambda,\lambda)\ C(LL'\nu;M-\mu+\lambda,$$

$$\times\ M'-\lambda)\ Y^{M+M'-\mu}_{\nu}\ \hat{\Sigma}_{\mu} \tag{70}$$

Vector identities useful in proving the above results:

$$\vec{\nabla} = \hat{r}(\hat{r}\cdot\vec{\nabla}) - \hat{r} \times (\hat{r} \times \vec{\nabla}) = \hat{r}\ \partial/\partial r - i(\hat{r} \times \hat{L})/r$$

$$\vec{\nabla} \times (\vec{r} \times \vec{\nabla}) = \vec{r}\ \nabla^2 - \vec{\nabla}(1 + r\ d/dr)$$

$$\vec{r} \cdot (\vec{\underset{\sim}{L}} \times \vec{A}) = 2i\ \hat{r} \cdot \vec{A} - \vec{\underset{\sim}{L}} \cdot (\hat{r} \times \vec{A})$$

$$\hat{r} \times (\vec{\underset{\sim}{L}} \times \vec{A}) = \vec{\underset{\sim}{L}}\ (\hat{r} \cdot \vec{A}) + i\ \hat{r} \times \vec{A}$$

$$\vec{\underset{\sim}{L}} \times (\vec{\underset{\sim}{L}} \times \vec{A}) = \vec{\underset{\sim}{L}}(\vec{L} \cdot \vec{A}) + i\ \vec{\underset{\sim}{L}} \times \vec{A} - \underset{\sim}{L}^2\vec{A}$$

$$\vec{Y}^{M}_{JL} \equiv \hat{r}\ (\hat{r} \cdot \vec{Y}^{M}_{JL}) - \hat{r} \times (\hat{r} \times \vec{Y}^{M}_{JL}) \tag{71}$$

$\vec{A}$ is any vector

$$\vec{\underset{\sim}{L}} = -\ i\vec{r} \times \vec{\nabla}$$

TABLES OF CLEBSCH-GORDAN COEFFICIENTS

$$C\,(L1J;M_L\;m\;M) \equiv C^{L\;1\;J}_{M_L\,m\,M}$$

J	m = 1	m = 0	m = -1
$L + 1$	$\left[\frac{(J+M-1)(J+M)}{(2J-1)(2J)}\right]^{\frac{1}{2}}$	$\left(\frac{(J-M)(J+M)}{(2J-1)(J)}\right)^{\frac{1}{2}}$	$\left(\frac{(J+M-1)(J-M)}{(2J-1)(2J)}\right)^{\frac{1}{2}}$
L	$-\left[\frac{(J+M)(J-M+1)}{2J(J+1)}\right]^{\frac{1}{2}}$	$\frac{M}{[J(J+1)]^{\frac{1}{2}}}$	$\left[\frac{(J-M)(J+M+1)}{2J(J+1)}\right]^{\frac{1}{2}}$
$L - 1$	$\left[\frac{(J-M+1)(J-M+2)}{2(J+1)(2J+3)}\right]^{\frac{1}{2}}$	$-\left[\frac{(J-M+1)(J+M+1)}{(J+1)(2J+3)}\right]^{\frac{1}{2}}$	$\left[\frac{(J+M+1)(J+M+2)}{(2J+2)(2J+3)}\right]^{\frac{1}{2}}$

$$C\,(L\tfrac{1}{2}\;J;M_L\;\tau\;M) \equiv C^{L\frac{1}{2}\;J}_{M_L\,\tau\,M}$$

J	$\tau = \frac{1}{2}$	$\tau = -\frac{1}{2}$
$L + \frac{1}{2}$	$\left[\frac{J+M}{2J}\right]^{\frac{1}{2}}$	$\left[\frac{J-M}{2J}\right]^{\frac{1}{2}}$
$L - \frac{1}{4}$	$-\left[\frac{J-M+1}{2J+2}\right]^{\frac{1}{2}}$	$\left[\frac{J+M+1}{2J+2}\right]^{\frac{1}{2}}$

REFERENCES

1. E. U. Condon and G. H. Shortley, "Theory of Atomic Spectra," Cambridge University Press, London, 1959, Chap. III.

2. M. E. Rose, "Angular Momentum," Wiley, New York, 1957, p. 76.

3. A. R. Edmonds, "Angular Momentum," Princeton University Press, Princeton, 1957.

4. J. D. Jackson, "Classical Electrodynamics," Wiley, New York, 1964, p. 538.

5. H. C. Corben and J. Schwinger, Phys. Rev., 58, 953 (1940).

6. W. W. Hansen, Phys. Rev., 47, 139 (1935).

7. H. Bethe and E. E. Salpeter, "Quantum Mechanics of One and Two Electron Atoms," Academic, New York, 1957, p. 344.

8. E. Merzbacher, "Quantum Mechanics," Wiley, New York, 1970, p. 185.

9. E. L. Hill, Am. J. Phys., 22, 211 (1954).

10. A. Erdelyi, "Higher Transcendental Functions," Vol. 2, Bateman Manuscript Project, McGraw-Hill, New York, 1953, p. 232.

11. M. E. Rose, "Multipole Fields," Wiley, New York, 1955, p. 10.

12. L. C. Biedenharn, lecture notes on multipole fields (unpublished).

13. J. M. Blatt and V. F. Weisskopf, "Theoretical Nuclear Physics," Wiley, New York, 1952, p. 796.

APPENDIX B

THE DIVAC δ FUNCTION

A quantity $\delta(x)$ depending on an independent variable x satisfying the conditions

$$\int_{-\infty}^{\infty} \delta(x)\ dx = 1$$

$$\delta(x) = 0, \qquad x \neq 0 \tag{72}$$

defines a one-dimensional δ function.

The following integral is sometimes regarded as an alternative definition

$$\int_{-\infty}^{\infty} f(x)\ \delta(x-a)\ dx = f(a) \tag{73}$$

where a is a real number.

Another way of defining the δ function is as the differential coefficient of the function (step function) given by

$$f(x) = 0 \qquad x < 0$$
$$= 1 \qquad x > 0$$

$$\frac{d}{dx} f(x) = \delta(x) \tag{74}$$

As is well known, the meaning of any of the following equations is that its two sides give equivalent results as factors in an integrand

(1) $\delta(-x) = \delta(x)$

(2) $x\delta(x) = 0$

(3) $\delta(a\,x) = (1/a)\delta(x) \qquad a > 0$

(4) $\delta(x^2 - a^2) = (1/2a)[\delta(x - a) + \delta(x + a)] \qquad a > 0$

more generally

$\delta((x - a)(x - b)) = (1/|a-b|)\ \{\delta(x - a) + \delta(x - b)\}$

(5) $\int \delta(a - x)\ \delta(x - b)\ dx = \delta(a - b)$

(6) $f(x)\ \delta(x - a) = f(a)\ \delta(x - a)$ (75)

It is interesting to note that

$(d/dx) \log x = (1/x) - i\pi\ \delta(x)$

Some "representations" of the δ function are given below.

(1) $\lim_{g \to \infty} \sin g\,x/\pi x = \delta(x)$ (76)

(2) $\frac{1}{2}\pi \int_{-\infty}^{\infty} e^{ikx}\, dx = \delta(k)$ and similarly

$\frac{1}{2}\pi \int_{-\infty}^{\infty} e^{ikx}\, dx = \delta(x)$ (77)

(3) (almost same as 2)

$\mathrm{Lim}_{L \to \infty} (\frac{1}{2}\pi) \int_{-L}^{L} \cos kx\, dk = \delta(x)$ (78)

(4) If $U_n(x)$ represents a complete set of ortho-normal functions, that is,

$\int_{\infty}^{\infty} U_n^*(x)\ U_n(x)\ dx = \delta_{mn}$

then

$$\sum_n U^*_n(x)\, U_n(x') = \delta(x - x') \tag{79}$$

$$(5)\quad \lim_{a\to 0} (1/(\pi a)^{1/2}) \exp(-(1/a)x^2) = \delta(x) \tag{80}$$

$$(6)\quad \lim_{a\to 0} a/[\pi(x^2 + a^2)] = \delta(x) \tag{81}$$

$$(7)\quad (1/2)(d^2/dx^2)|x| = \delta(x) \tag{82}$$

Notice that Equation (75) shows that the δ function is an even function (even 'parity').

$$(8)\quad \delta'(x) \equiv (d/dx)\,\delta(x) = -(1/x)\,\delta(x)$$

$$(9)\quad \delta(x - a) = \lim_{a\to 0} (1/\pi)\,\{a/[a^2 + (x - a)^2]\}$$

$$(10)\quad \lim_{\lambda\to\infty} (\lambda/\sqrt{\pi})\exp(-\lambda^2 x^2) = \delta(x)$$

$$(11)\quad \delta[(1/x) - (1/y)] = xy\,\delta(x-y) = x^2\delta(x - y)$$

$$= y^2\delta(x - y)$$

$$(12)\quad \delta(f(x)) = \sum_n \{\delta(x - x_n)/|f'(x_n)|\} \tag{83}$$

where $f'(x) \equiv (d/dx)f(x)$ and $f(x_n) = 0$

THREE-DIMENSIONAL δ-FUNCTIONS (IN ORTHOGONAL COORDINATES)

$$\delta(\vec{r} - \vec{r}\,') \equiv \delta(x - x')\,\delta(y - y')\,\delta(z - z')$$

$$\int \delta(\vec{r} - \vec{r}\,')\, d\vec{r}\,' = 1, \quad d\vec{r} \equiv dx\, dy\, dz \tag{84}$$

(1) If $U_E(\vec{r})$ represent a complete orthonormal set of functions, that is,

$$\int U^*_{E'}(\vec{r})\ U_E(\vec{r})\ d\vec{r} \ = \ \delta_{EE'}$$

then

$$\sum_E U^*_E(\vec{r}')\ U_E(\vec{r}) \ = \ \delta(\vec{r} - \vec{r}') \tag{85}$$

for instance,

$$[1/(2\pi)^{3/2}] \int e^{i\vec{k}\cdot(\vec{r}-\vec{r}')}\ d\vec{k} = \delta(\vec{r} - \vec{r}') \tag{86}$$

$$\nabla^2 \equiv \partial^2/\partial x^2 + \partial^2/\partial y^2 + \partial^2/\partial z^2, |\vec{r} - \vec{r}'| = \{(x - x')^2 + (y - y')^2 + (z - z')^2\}^{1/2}$$

(2) $$\nabla^2(1/|\vec{r} - \vec{r}'|) = -4\pi\ \delta(\vec{r} - \vec{r}') = \nabla'^2(1/|\vec{r} - \vec{r}'|) \tag{87}$$

(3) $$\sum_{\ell=0}^{\infty} \sum_{m=-\ell}^{\ell} Y^{m*}_{\ell}(\theta'\psi')\ Y^m_{\ell}(\theta\psi) \ = (1/\sin\theta')\ \delta(\theta-\theta')\delta(\psi-\psi') = \delta(\cos\theta - \cos\theta')\ \delta(\psi-\psi')$$

$$\begin{aligned}\delta(\vec{r} - \vec{r}') &= (1/r'^2)\ \delta(r - r')\ \delta(\cos\theta - \cos\theta')\ \delta(\psi-\psi') \\ &= (1/r'^2 \sin\theta')\ \delta(r - r')\ \delta(\theta - \theta')\ \delta(\psi - \psi') \\ &= (1/\rho')\ \delta(\rho - \rho')\ \delta(\psi - \psi')\ \delta(z - z')\end{aligned} \tag{88}$$

In generalized (orthogonal) coordinates, if

$$ds^2 = dx^2 + dy^2 + dz^2 = h_1^2 \, d\,\xi_1^2 + h_2^2 \, d\,\xi_2^2 + h_3^2 \, d\,\xi_3^2$$

$$\delta(\vec{r} - \vec{r}') = \delta(\xi_1 - \xi_1') \, \delta(\xi_2 - \xi_2') \, \delta(\xi_3 - \xi_3')/h_1 h_2 h_3 \quad (89)$$

REFERENCES

1. P. M. Morse and H. Feshbach, "Methods of Theoretical Physics," Part I, McGraw-Hill, New York, 1953.

2. J. R. MacDonald and M. K. Brachman, Rev. Mod. Phys., 28, 414 (1956).

3. P. A. M. Dirac, "Quantum Mechanics," Oxford University Press, Oxford, 1947.

4. L. I. Schiff, "Quantum Mechanics," McGraw-Hill, New York, 1969.

5. J. D. Jackson, "Classical Electrodynamics," Wiley, New York, 1962.

6. M. J. Lighthill, "Introduction to Fourier Analysis and Generalized Functions," Cambridge University Press, Cambridge, 1958.

INDEX